Search for Universal Ancestors

The Origins of Life

Editors:

H. Hartman

Massachusetts Institute of Technology
Cambridge, Massachusetts

J. G. Lawless

NASA Ames Research Center
Moffett Field, California

P. Morrison

Massachusetts Institute of Technology
Cambridge, Massachusetts

Originally Prepared at Ames Research Center

BLACKWELL SCIENTIFIC PUBLICATIONS

Palo Alto   Oxford   London   Edinburgh   Boston   Melbourne

Editorial Offices

667 Lytton Avenue, Palo Alto, California 94301
Osney Mead, Oxford, OX2 0EL, UK
8 John Street, London WC1N 2ES, UK
23 Ainslie Place, Edinburgh, EH3 6AJ, UK
52 Beacon Street, Boston, Massachusetts 02108
107 Barry Street, Carlton, Victoria 3053, Australia

Library of Congress Cataloging in Publication Data

Search for the universal ancestors.
    Includes index.
    1. Life-Origin. 2. Evolution. 3. Solar system—
Origin. I. Hartman, H. (Horst), 1937—
II. Lawless, J. G. (James G.) III. Morrison, Philip.
QH325.S42   1987   577   86-31685
ISBN 0-86542-328-8

Cover design and photograph: Gary Head

Distributors

USA and Canada
Blackwell Scientific Publications
P.O. Box 50009
Palo Alto, California 94303

Australia
Blackwell Scientific Publications (Australia) Pty Ltd
107 Barry Street, Carlton
Victoria 3053

UK
Blackwell Scientific Publications
Osney Mead
Oxford OX2 0EL

TABLE OF CONTENTS

LIST OF CONTRIBUTORS

M. Calvin, University of California, Berkeley

S. Chang, NASA Ames Research Center

D. Deamer, University of California, Davis

A. Delsemme, University of Toledo

J. Ferris, Rensselaer Polytechnic Institute, New York

C. Folsome, University of Hawaii

H. Hartman, Massachusetts Institute of Technology

J. Hayes, Indiana University

H. Holland, Harvard University

N. Horowitz, California Institute of Technology

J. Lawless, NASA Ames Research Center

L. Margulis, Boston University

S. Miller, University of California, San Diego

P. Morrison, Massachusetts Institute of Technology

B. Nagy, University of Arizona

P. O'Hara, NASA Ames Research Center

L. Orgel, Salk Institute

J. Oro, University of Houston

T. Owen, State University of New York at Stonybrook

C. Ponnamperuma, University of Maryland

A. Rich, Massachusetts Institute of Technology

J. W. Schopf, University of California, Los Angeles

W. Stoeckinius, University of California,
San Francisco Medical Center

D. Usher, Cornell University

G. Wetherill, Carnegie Institution of Washington

D. White, University of Santa Clara

C. Woese, University of Illinois, Urbana

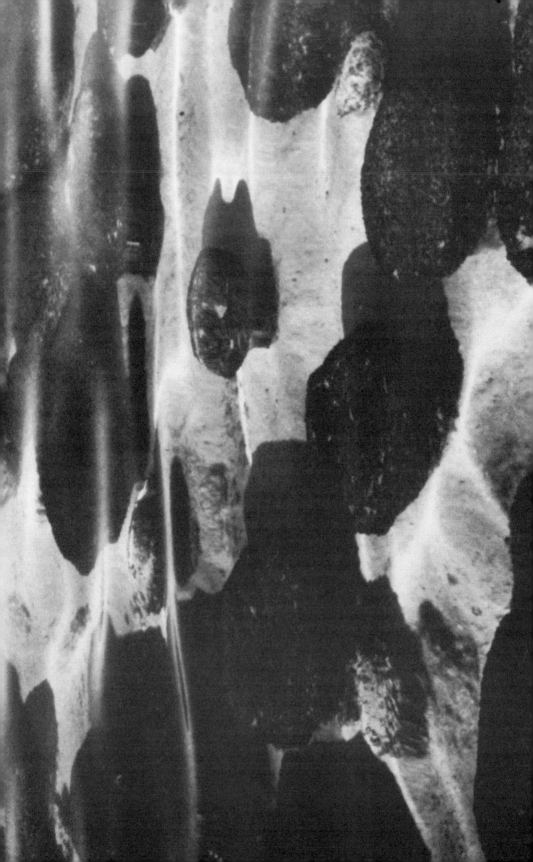

PROLOGUE

The day was calm. The waves that lapped the shore were small, almost lazy; they were the waves of shallow waters, not of the open ocean. A ruddy sun shone in the hazy sky. The slow stream that came down to the foreshore rippled a little in the light wind, and the pebbles tumbled here and there without much energy. Drama was not wholly absent, for along the skyline all but lost in the distance were two or three volcanic cones. They were quiet just now, but a walk along the beach would soon bring a traveler to a stretch of all but impassable lava, where once, not so long back, the molten rock had oozed and hissed into the waters from an inland fissure. It would happen again, but no one could foresee just where and when the encounter would take place.

The day was calm and the scene was lonely. The beach was devoid of shells. No flies buzzed; nothing at all hopped or crawled along the water's edge. No birds flew; no fish swam in the sea; no clawed creatures scuttled below the tidal waters. The rocky lands inward from the sea were utterly barren of life. Neither lizards nor mice could be found, and neither a tree nor a blade of grass spread green blades to the sunshine. Yet life was present, even abundant, in the scene. It grew everywhere that the shallow waters brimmed out to dry land: dense knobs and sheets of algae and bacteria covered all the shallows, out into the bay and up the stream toward the higher lands. That life was never out of touch with water; it never survives higher than a matter of inches from moisture. Inland, here and there, a few dry old knobs could be found, quite whitened, rocklike — a growing mat of the only life in this quiet land, stranded forever by some shift in the watercourse.

Modern structures formed by microbial mats found in Australia are analogous to structures formed by ancient microbial mats billions of years ago in a similar environment.
Photo courtesy of J. W. Schopf

Just such a scene — we can infer the details rather well from the complex fabric of the rock samples — would present itself at the spot, now the western coast of Australia, where the oldest trace of life in all the Earth is found. The time is long ago indeed, a time we can estimate to within a few percent from a secure, mutually confirming set of radioactive decay measurements. The signs of copious algal life, which bears a remarkable resemblance to the same forms found throughout the record of the rocks up to the present day, occur almost as early as the first dated rocks. One must emphasize that this teeming life, single-celled, though colonial in nature, was about all that lived on Earth, not only for the first pages of the record but for four-fifths of our whole past. Not until a time only 0.7 billion years (b.y.) back can we surely see any relic of life more mobile than the algal and bacterial mats. Indeed, they themselves become more complex in microstructure and more powerful in their chemistry over the 3 b.y. of their evolution and change. No life is mobile (beyond the drift of plankton) until about that time, 0.6 or 0.8 b.y. ago. And it requires another couple of hundred million years before anything alive, either plant or animal, can break close contact with the waters, to stand well above the coast, the marsh, or the damp soil. All the forms of life we see in the familiar fossil record, everything so graphically drawn by the paleontological artists, all those feathery sea lilies, bulge-eyed trilobites, all sharks and dinosaurs, all ancient birds and beasts, all dawn palms or big kelp, all that crawls and flies and swims and stalks, all that branches or flutters in the wind, belong to the last 10% or 15% of life's long history on Earth.

The direct rock record supports one major plausible inference: All life we know has evolved from single, small-celled beginnings, forms like those still to be found as copious and vigorous participants in the great geochemical cycles, the blue-green algae and their kin, the bacteria. These had filled the shallow-water world in full vigor even by the time of our earliest evidence. Theirs, of course, is the triumph of lowly biochemistry — not motion, not sensory response, not even structure on the scale of naked-eye visibility. It is rather the microstructure, the complexities at the level of molecular helixes, sheets, tubes and rods, and the complex biochemical pathways they enable, which evolved over the entire first half or so of life's biography. Thus, it is biochemistry we must search out back to the time of the most ancient rocks known, and before, if we can. The winds and the waters, the volcanic fires and the slumping sands, the lava pours and the rolling stones — those were not much different from today. But in fact we do not even know whether human beings — time travelers — could have breathed that otherwise commonplace breeze. Was it oxygenated? Could the old algal mats already use oxygen and sunlight to build their substance in the open air? Or had they not yet made that invention, so that they subsisted in a very different atmosphere from ours, perhaps not yet even making good use of solar photons? We cannot be sure. We know that many bacteria today are fit for vigorous life oxygen-free. We know, moreover, that all the oxygen of today's atmosphere is turned over very rapidly by the green-plant world. But just when that ability first arose has not yet been fixed. It is in fact the molecular facts we still seek, with much difficulty, there in the most ancient rocks. For it is on that level that the mechanisms of life must have begun, about 4 b.y. back. But the rocks are steadily reworked, buried, heated, and reheated in the Earth's fiery mantle. We have to look hard indeed to find rocks older than Isua in Greenland, which may — or well may not — bear the crucial evidence we seek. Certainly, that search will go on, and with sharper analyses we will find more clues in the ancient, disturbed, and enigmatic samples.

THE DEPTH OF SPACE

Perhaps, we can work our way forward in time to life, forward from the date when the Sun and its planets were somehow condensed jointly out of the interstellar gases. We know that date rather well. The surface of the Moon, a large collection of meteorites that have been trapped from orbit by their chance encounters with Earth, and the strongly supported inferences from the dating of long series of Earth rocks, all lead to the conclusion that our planet was assembled by a complex set of processes just about 4.5 (±0.1) b.y. ago. In the time between 4.5 and 3.8 b.y., we know Earth became the round sphere it is, the rocks of the crust grew solid and were well sorted, and the processes of ordinary geology approached the familiar. This stormy period, with an epidemic of heavy impacts on planetary surfaces, is under detailed study which includes the hints given by the other planets and satellites whose compositions we come more and more to know. The meteors, too, offer samples stored in the refrigerator of orbit by which we try to conclude what material made up the Earth. We are pretty sure that there was plenty of time, for the orbital processes are speedy compared to most geological ones. Free fall is faster than the drift of continents and the weathering of mountains to fill the seas with silt. Even such grand geological dramas take a fifth of a billion years at most. So there is ample time; the physical processes were more or less over and became familiar within a couple of hundred million years (m.y.) or so. But what that earliest stable Earth was like is a question still beyond the grasp of our models; the biologists look to the planetary sciences for the initial conditions of sea and air, and the physical studies still seek a constraining hint from what the biochemists will tell them of the initial atmosphere!

There has been one unexpected finding in space within the last decade especially. It is the universal prevalence of carbon-containing molecules. The linked atoms of carbon were once thought to be solely the work of life. Now we find them overall, in plenty. A substance like ethyl alcohol exists in vast quantities, though gaseous and dilute to the point of vacuum, spread in the huge, thin clouds of interstellar gas. Substances of even more complex and life-like kind are found in the meteorites that fall now and again to Earth, some of which resemble a blackened cheese, soft and carbon-rich in substance. These are certainly products of the evolution of the meteorites, perhaps somewhat associated with the debris of minor planets or at least with the conditions of the solar nebula where the planets grew. There is a chance that these carbon compounds may have contributed such useful organic resources directly to the early Earth, which was certainly heavily bombarded with gifts from orbit. Most workers believe that the surface of the Earth itself was fully suited for the production of organic carbon compounds from the commonplace atoms, like carbon and hydrogen, which were already in place. It is the density which must be of importance, for complex molecules arise out of the close proximity of several atoms leading to their eventual union, an event none too common within the dilution of space. Exactly that condition (the temperature, too, must be permissive) gives our Earth its liquid water, intimately, indispensably part of life, ancient and modern. Perhaps no other place we have yet found in the cosmos allows the presence of liquid water. Mars long ago had great rivers, we think. Water and carbon-containing molecules fit for living forms go tightly together. The conclusion which everyone accepts is that small carbon-containing molecules form spontaneously, under the right ambient conditions, whether or not life is directly involved. If not the true precursors of life, these molecules are at least patterns for the loom of life. They can be made by a variety of synthetic processes; they differ indeed only in detail from nonorganic molecules, copious but biologically insignificant in the atmospheres of the cooler stars. The molecule is plainly implicit in the atomic world.

We will press the rock record back to its first pages. We will improve the planetary and meteorite studies forward to the Earth in birth. But can we look squarely into that half-billion year gap between the birth of the planet and the oldest relics of life?

THE ESSENTIAL MEMORIES

The extension of 35 years of molecular biology involving the idea of the genetic code and its work, has shown us much of the inner nature of life, especially of microbial life. The effort to frame a general definition of life, which might seem at the core of the matter, is in fact not so salient in the search for life's origin. For what we seek must be the long chain of events which gave rise to that specific complex web of present-day life on Earth of which we are a part. Even if quite other forms are possible, they would seem less relevant to the quest for what happened here. It is no surprise, then, that the key questions have been sought on theoretical grounds. For 20 years people have looked for conditions which, under the little-known natural circumstances of 4 b.y. ago, could plausibly lead to the rise of proteins — the working molecules of life which marshal the linked chemical reactions to build living systems and lend them function. Parallel effort has been spent with the nucleic acids, which on Earth alone form those subtle, long molecules (DNA and RNA), as well as the various copying peripherals that embody and reproduce the instructions for the proteins which implement the order, even the act of code reproduction itself. These complex molecules are universal in life forms today; they have long seemed a wise starting point for the search for origins.

Neither a coded message, whose slow elaboration is the only key to the evolutionary path, nor these protein jigs and fixtures, which alone allow the expression of the inert code within the world of living change, can be the whole story. No code, no way to elaborate the chain of life forever; but no jigs, no action on the world. A biological contract between these molecules, or more strictly, between these two functions in some molecular form, seems the center of the issue. We do not see it clearly yet. The clues from present life are useful, but today the mechanism has become so precise, so well-functioning, so long-elaborated that the steps that led to it from a simpler, nonliving world are hard to guess. It is possible that some simplified intermediate structures will be found; it is also possible that some key contrivance, now entirely superseded, must be found. Some think that the inorganic substructure of some mineral might have offered a crystalline framework for the first systems of molecular self-reproduction. Other essentials besides the message and the action are postulated; or perhaps some early form of the message may have been at least weakly self-acting. The whole topic is complex, central, challenging.

One conclusion seems stronger than ever, touching on the eons of evolutionary elaboration. It seems likely that the first cells of the microbial mat were the unusually small and structurally more or less simple ones which still distinguish the bacteria and the blue-green algae as a group from all other higher forms of life. Sometime in the first 2 b.y., new cell types arose. More or less algal, they were big, hundreds of times the volume of their forebears, like most cells of the animals and plants of today. Moreover, they held the specialized organelles of cells, those which today carry out efficient photosynthesis, hold and take out the message molecules safely and precisely, arrange for sequential reactions of energy storage and liberation, etc., in all higher forms. The smaller cells perform the same functions (indeed, as biochemists they are even more versatile), but they lack many advantages of rate and of diversity in reproduction. There is good reason to believe that here we see another later social contract – the organelles of the larger cells are perhaps the offspring of once-independent organisms, which long ago contracted to dwell within some host cell and share its world. A few such symbiotic arrangements may have made the complex and protean cells of today, the eukaryotes. Their cooperation to form organelles, then whole creatures, has given rise to the multicellular forms of life that are now large enough to make an imprint in the world one by one, and enterprising enough to add swift mobility and varied macrostructure to the chemical virtuosity of that ancient algal mat.

Probably, these several contractual unions are the heart of our problem; so far only the rise of the organelles seems to find some support in the world of life as we study it today.

THE TENSIONS OF RESEARCH

There is a certain tension in scientific research. No doubt all researchers would like to solve important problems, problems with impact on the mind, or problems whose solutions bear on human needs in a material way. But those problems are hard, often refractory. At any state of knowledge, the researcher is therefore led to seek out, not the most important problems, but the soluble ones. Galileo watched the lamp swing with uniform beat; he could get somewhere with that. But his early effort to explain the tides was not the source of much further work. In the study of the origin of life some balance must be struck. The problem is clearly important; perhaps no other problem speaks more to the common concern of all reflective human beings to learn our place in the world. But it is evidently no simple problem, more so because it cannot be sought within a single discipline. Here, molecular biology meets astronomy, both encounter geology, and each of these major disciplines draws upon the chief results of chemistry and physics over a very wide range. Such a specialty is institutionally fragile; its devotees generally must find their professional niches in quite different, better-focused enterprises — those with degrees, standard texts, students, clear applications. We believe that the maturity and importance of the problem begin to demand explicit support and recognition. The question lies squarely across the major currents of microbiological and of planetary research today. Those sciences could reach no more important outcome than to illuminate the origins of life on Earth.

In this search, we therefore offer an account of the state of our knowledge, our hopes, and our puzzles as they are today. We hope the material will convince the reader how fascinating and how timely is this topic in the present state of science. We might remark also that a focus on the widespread but rather quiet, cryptic life of the microbial mats of the great watery flats is strangely relevant to today's world in which the geochemical and atmospheric cycles themselves begin to show the effects of human intervention for good and bad. The great role of that lowly life in the deep past is to a degree continued today; we know all too little of its complex and rich operations within the changing balance of contemporary nature. As we look back into the past, we will both employ and illuminate the present. But above all, what we seek is not only practical advantage, though that will come. Our chief goal is a kind of self-knowledge, as deep as our oldest myth: how it came about on this Earth that the quick were first parted from the dead.

Philip Morrison
Massachusetts Institute of Technology
Cambridge, Massachusetts

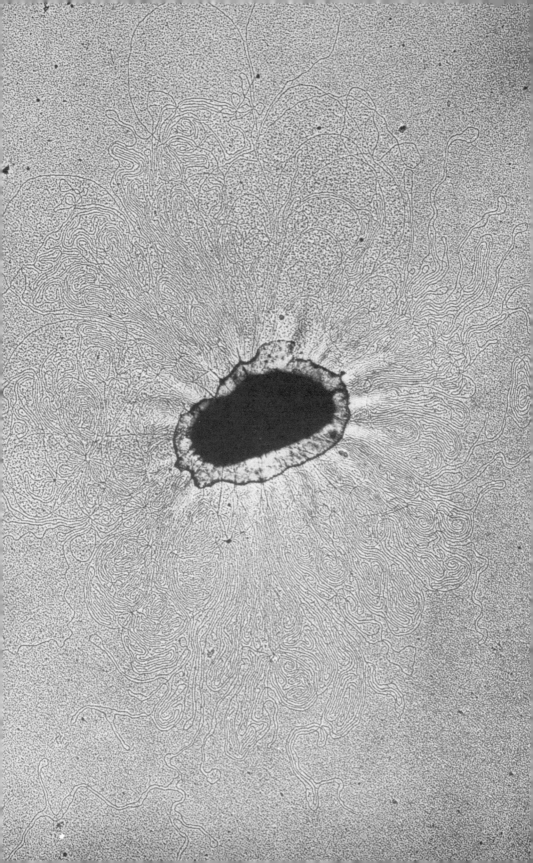

I. TOWARD THE ORIGINS OF LIFE

A DEFINITION OF LIFE?

The search for a clear definition of life is in itself a scientific problem of real depth. In the present context, that search appears a little peripheral, however deep and urgent the issue may be to a full understanding of biology. For our task is not at all to examine every path that the lifelike might walk. The classical metaphors of the flame, the whirlpool, the growth of a crystal, even so speculative a diversion as the notion of life based on silicon, whether within some computer mainframe or in the creaking motion on some far-off planet, are somewhat beside the point. The task is set for us sharply by the historical event: The life whose beginnings we seek is that of our own planet, carbon-based, molecularly elaborated, and capable of evolving into the web of life shown in the geological record up to our own time.

The attributes by which we recognize living things as alive — their capacity to grow, replicate, and repair themselves, to produce elaborate and seemingly purposeful structures and behaviors, to adapt to the most varied conditions — are derived ultimately from the genetic properties of living matter. By genetic properties we mean these two: self-duplication and discrete change. That is to say, living systems are systems that reproduce themselves closely, but that mutate as well, and then can reproduce their mutations. These properties define the living state. Such systems, mutating — albeit blindly — in many directions, will evolve through the process of natural selection. In time, they can yield the living world in its endless variety and complexity.

Within the notion of self-duplication, so intrinsic to life, there once seemed to rest a logical disaster. A mechanism which must reproduce itself needs to be complicated. But the more complicated

Electron micrograph of DNA by Lorne MacHattie, in collaboration with Ken Berns and Charles A. Thomas, Jr.

it is, the harder it is to reproduce, and thus the more complicated it must still be — that particular kind of trouble was once and for all eliminated by the arguments of the mathematician John von Neumann 30 years ago. His argument was well-summarized by Freeman Dyson:

"Von Neumann did not live long enough to bring his theory of automata into existence. He did live long enough to see his insight into the functioning of living organisms brilliantly confirmed by the biologists. The main theme of his 1948 lecture is an abstract analysis of the structure of an automaton which is of sufficient complexity to have the power of reproducing itself. He shows that a self-reproducing automaton must have four separate components with the following functions. Component A is an automatic factory, an automaton which collects raw materials and processes them into an output specified by a written instruction which must be supplied from the outside. Component B is a duplicator, an automaton which takes a written instruction and copies it. Component C is a controller, an automaton which is hooked up to both A and B. When C is given an instruction, it first passes the instruction to B for duplication, then passes it to A for action, and finally supplies the copied instruction to the output of A while keeping the original for itself. Component D is a written instruction containing the complete specifications which cause A to manufacture the combined system A plus B plus C. Von Neumann's analysis showed that a structure of this kind was logically necessary and that it must also exist in living cells. Five years later Crick and Watson discovered the structure of DNA, and now every child learns in high school the biological identification of Von Neumann's four components. D is the genetic material, DNA; A is the ribosomes; B is the enzymes RNA and DNA polymerase; and C is the repressor and derepressor control molecules and other items whose functioning is

still imperfectly understood. So far as we know, the
basic design of every micro-organism larger than a
virus is precisely as Von Neumann said it should be."

This qualitative triumph was not the last of the mathematical
approaches. Perhaps the most sophisticated work which followed was
that of the seventies by the distinguished physical chemist, Manfred
Eigen. He and his colleagues have elaborated a carefully controlled
model of linked chemical reactions, which shows the possibility of
the steady increase of chemical complexity by various feedback
mechanisms entirely within their model domain of interdependent
reaction rates and yields: the explicit working out of the vague old
notion of autocatalysis. But it seems rather far from our concrete
problems, for it begins with a mechanism of reproduction with
interaction. Yet it is the origin of that mechanism which is a central
problem.

Once again, the impact of this abstract work is, so far, stronger
at the level of general understanding than upon the actual search for
terrestrial beginnings; most workers prefer to take strong clues from
the specificity of the life we know than to follow logical consequences of plausible general models. There are hints of experimental
systems: Certain bacterial cells, products of some error in cell division, metabolically complete but without genetic apparatus, might
allow a kind of direct experiment in the domain of self-reproducing
automata. But most effort still seems directed to a more chemically
characterized level.[1] For most work so far, that has been the pursuit
of a bridge across the present wide gulf between the stable nucleotide
memory string – the DNA – and the reactive peptides of enzymes,
the structural proteins, and perhaps some other membrane constituents. The emergence of a self-reproducing system from single molecules is yet elusive.

LIFE ON A TAPE

Perhaps the most important achievement of modern biology has
been the discovery of the chemical structures and mechanisms

---

[1] Important representative chemicals are illustrated in the appendix.

responsible for the genetic properties of living things, the components A through D of von Neumann's automaton. These properties derive from just two classes of large, information-rich molecules: proteins and nucleic acids. The nucleic acids are the ultimate self-replicating, stable, yet mutable structures of all living matter today. They form the genes, the bearers of the genetic heritage, in every known species. This heritage consists of durable information, largely concerned, directly or indirectly, with the production of specific protein molecules. The latter form most of the cellular structures and the chemically active enzymes — the versatile class of highly efficient, interactive catalysts that control the chemical activities of cells, including the eventual self-synthesis of more enzymes, other proteins, nucleic acids, and other key molecules. The nucleic acids and proteins thus constitute an interlocking and interdependent association, the genetic system. Whatever is unique about living matter is inherent in this system. There is an instructing tape, the nucleic acid, which directs the assembly of the universal chemical tools, the enzymes. Note, though, that all the important structures are not single molecules, but complexes like lipid membranes.

The duality of the genetic system arises from the circumstance that survival in the struggle for existence depends on the ability of organisms to synthesize a large variety of specific proteins; those proteins are highly ordered, hence, thermally improbable structures that must be built up by a long sequence of individual amino acids. If every generation had to discover for itself how to assemble amino acids in the right order to produce the proteins it needs, survival would be impossible. This information must be transmitted from parent to offspring, and a mechanism for storing and copying it is required. Amino acid sequences cannot be copied by any known chemical scheme from a preexisting protein, but nucleotide sequences can be copied from a nucleic acid. Consequently, the instructions for assembling protein molecules are encoded in nucleic acids. Only the latter are copied for inheritance; only the former are made anew to do the work.

As indicated above, the lengthy information sequence contained in the genes was generated by random mutations in DNA, screened by natural selection. The genetic specifications are thus the evolutionary product, a record of discovered solutions to the problems of survival encountered by the species in the long course of its history.

Basic to this evolution is the mechanism — itself slowly perfected by mutation and natural selection — that brings about the nearly flawless replication of DNA and its mutants, and through them, of proteins and the entire organism.

The spontaneous origin of so complex a system as that described above poses great conceptual difficulties. It must be recognized that the simplest modern genetic system we see is highly evolved; the original system was probably much simpler. Two logically possible predecessors are worth exploring: (1) polynucleotides which still retain some catalytic capability, and (2) polypeptides which yet have some replicative capability. Molecular systems such as these would be extremely inefficient by standards of life today, but they might have been sufficiently accurate under the less-competitive conditions of the primordial Earth to survive and to evolve. Either one might have been capable of developing into the modern dual mechanism.

THE LOCALIZATION OF ORDER

The search for interactions among molecules which might lead to self-replicating systems of either or both kinds of molecules, nucleic acids and proteins, has progressed to some depth. Biochemical preparations which are essentially homogeneous solutions can yield remarkable results whenever the polymer prototypes required are present in rather high concentration. The classical experiments on DNA replication *in vitro* are only one famous example.

But this sort of system is not without its limitations. It is after all the cell, a large set of interactive molecules, which is the unit of present-day life. The notion that simple mutations — if one thinks of many successive single-point modifications within a long polynucleotide sequence — could in the fullness of time lead to the marvelous outcomes of natural selection is inherent in the point of view. Surely, that is too simple a view of the actual course of evolution. The time required for a given selective result can be much reduced if the process proceeded among quite distinct strings of genes, to join several of them later in an action which may confer complex, new capabilities all at once. This is of course seen in living forms at many genetic levels, from that of the bacterial plasmids to sexual recombination in the classical way; even the typical microbial cell may itself largely be a symbiotic union of once-independent organisms. Some such hierarchical architecture is probably essential, even a rare process which increases the probability that successful changes can come to dominate. The slow growth of molecular tapes in solution — whether one imagines an organic ocean or a more plausible multitude of enriched tide pools — is bound to be speeded up by any process which tends to sequester reactants in the right way.

The logical simplicity of the replicating single molecule is evident. On the other hand, the complex system we now see requires the participation, even *in vitro,* of an energy source and a number of specific auxiliary enzymes. In the cell of course much more is invariably present, from the multimolecular mechanism of the transcription apparatus to the enclosing membranes which maintain the concentrations and the integrity of every cell. The molecular coupling described above is a minimal early step.

It is plausible to extend the idea of the single molecule. The simplest extension is to a system of a few molecules, which seems closer to the working system of life today. The difficulty is still in the coupling within any simple molecular system.

The main logical point seems to be to extend the participation of the environment, in order to restrict the replicating event to a single protogene itself. To this end the free energy is thought to be provided by a medium rich in needed building blocks, stored there by nonbiological processes that produced the molecules required. It is not difficult either to imagine the presence of some nonspecific catalyst, say, a metal ion, which might assist the coupling sought between the major polymers.

The next step beyond the single molecule self-copying in an organically rich and slightly metallized medium is the idea of a nonspecific substrate, some solid mineral surfaces where adsorption might locally increase concentration, provide nonspecific catalysis, and allow the use of rare components. At this point of hierarchical steps of unification, perhaps on the model of the simplest one, if A can make not itself but can make B, while B in turn makes A, the single union is self-duplicating. Once the possible role of substrate is admitted, the scheme can go even further. We shall discuss in more specific terms in chapter VI the conjectures that involve substrate to the extent that a replication of pattern is itself seen as beginning first on a mineral surface.

None of these schemes is yet much supported by experiment or quantitative theory, but given that life arose in an abiotic environment, some stable and yet not immutable spatial ordering was a key part of the process. Was that ordering all spatial, structural, in the sense of the formation of discrete phases beyond the molecular scale? Or was it in part temporal, kinetic in the sense of variable reaction rates among molecules? Those rates could be self-controlled by catalytic feedback loops, as well perhaps as by temporal, even cyclic, changes in the external chemical environment, e.g., dry-wet, or light-dark. That both possibilities might have been of importance is not to be overlooked. In the present state of our knowledge we would hope for small steps along any of these paths.

THE CHEMISTRY OF LIFE: WHY CARBON? WHAT ELSE?

The capability for generating, storing, replicating, and finally utilizing large amounts of information implies an underlying molecular complexity that is known only among the compounds of carbon. The special properties of the carbon atom that make it suitable for the construction of large, complex, three-dimensional molecules which are, in fact, thermodynamically unstable but kinetically metastable, are discussed in textbooks of organic chemistry. While living forms contain significant amounts of hydrogen and oxygen (in the form of H_2O), no other element enters as many and complex compounds as carbon. As every student knows, there are more compounds of carbon known than there are of all the other elements put together.

Table I-1 lists the relative number of atoms of the chemical elements which comprise the particular sample of life described. The data give in round numbers the number of atoms for each important element (not the weight) among 100 atoms of the sample. Note small round-off errors.

Life on our planet is carbon-based despite the fact that carbon is a minor element in the Earth's crust (about 0.5% by weight). This

TABLE I-1.– SOME ELEMENTARY RECIPES FOR LIFE FORMS

Life form	Element							
	H	C	O	N	P	S	Total	Other
Key biological polymers								
DNA	38	29	19	11	3	---	100	
Typical protein	50	31	10	8.5	---	0.5	100	
Whole living forms								
Green plant (alfalfa)	57	6	33	4	0.2	0.03	100.3	Ca, 0.1; all others less than S
Animal with bony skeleton (human)	62	9.5	26	1.3	.3	.08	100.3	Ca, 1.3; K, 0.09; Na, 0.07; all others less than S

is comprehensible once the special properties of the carbon atom referred to above are considered. It becomes almost expected in the light of recent discoveries showing that carbon compounds complex enough to have biological importance are reasonably abundant in carbon-containing meteorites (carbonaceous chondrites), even in vast, cold, and dilute interstellar clouds. Analysis of carbonaceous chondrites has revealed the presence of numerous amino acids, including at least eight of the amino acids of living proteins. It has been generally concluded that these amino acids are of extraterrestrial and nonbiological origin. Equally remarkable is the demonstration, by microwave spectroscopy, of a variety of organic compounds in interstellar space, in association with dust clouds rich in molecular hydrogen. Among the substances identified in these clouds are intermediates familiar in the synthesis of amino acids and of purines, pyrimidines, and sugars – in short, precursors of the genetic system, albeit very dilute within those enormous astronomical volumes.

It is clear from these discoveries that nonbiological reactions leading to the formation of biologically interesting molecules have occurred and are still occurring in the universe on a grand scale. This suggests that wherever life may be found it will be carbon-based, not greatly different in chemistry from our own.

The cosmic abundance of atoms tends to fall steadily with atomic number. None of the key atoms in the biopolymers of early life is heavier than sulfur, atomic number 16. Living forms do make some use of a good many elements in addition to the major constituents of their biopolymers. Every shelly or bony creature uses calcium, while some forms have skeletal frameworks of silica; these are specialized structures, but they all use elements rather common in the surface minerals of Earth. Among the most abundant of minor atoms are the volatile elements of the mineral world which are easily outgassed, dissolved by water, and weathered out of the rocks to salt the sea and make up the fundamental electrolyte solutions within all living cells. These are magnesium, sodium, potassium, calcium, and chlorine (chlorine is in fact rarer than the others listed). Next most important is iron, a rather heavy atom (atomic number 26), which, because of its intrinsic nuclear stability, is unusually abundant in the cosmos for its weight. The iron atom plays a central role in life today within a number of indispensable metal-containing organic structures, the blood-red pigments of course, but others as well. Similar roles can be played by less common metal atoms, e.g., copper, cobalt, zinc, manganese, even molybdenum, and vanadium. These are by no means common elements, but their exploitation by life seems to be an opportunism of natural selection, making use even of a trace of some rare atom to take advantage of its special properties. The secondary elements in life, in addition to hydrogen (H), carbon (C), oxygen (O), nitrogen (N), phosphorus (P), and sulfur (S), amount at most to one or two atoms in a hundred; no living species is known to require any element heavier than iodine (atomic number 53). The vital iron atom is the most abundant heavier atom in the human body, yet there is in the body only about one iron atom for every 15,000 of carbon! In all of biology only humans make any use of the rare heavy metals, like lead, gold, and uranium; that use is not biological, but cultural.

WATER

The two most abundant compound-forming atoms in the cosmos are hydrogen and oxygen. Their most familiar compound, water, is widespread. But it is grains of solid ice and very dilute water vapor that we find copiously in space. The familiar liquid water is not known to us with certainty anywhere in the universe save on Earth. We suspect it was in the bodies where the carbonaceous meteorites formed, probably some class of asteroids. We might have seen signs of it in the margins of the Martian polar cap; maybe it is present underground on Mars. The point, of course, is that liquid water is a fleeting substance; it can persist only within a limited range of temperature at reasonably high pressure. Such a regime is from a cosmic viewpoint intermediate in temperature; 10 or 20 times above the temperature of the cold gases of space, but 10 or 20 times below the temperature of a star surface. To realize liquid water, the pressure must stay rather high, compatible almost certainly only with the surface gravity of a modest, cool, planet-sized body. Thus, liquid water and life as well seem to be phenomena of high density. The near-vacuum of space cannot keep liquid water, and the atomic collisions which allow sequential reactions there are haltingly infrequent, even though quite numerous within the clouds in space. So familiar an "organic" compound as ethyl alcohol is well detected in such interstellar clouds. The total amount present there is huge; a single cloud contains more alcohol than all life on Earth has made over all its history, but it remains more dilute than a laboratory vacuum. It is so dilute there that any buildup to truly complex molecules is painfully slow.

Water, the medium of life, dominates life today. Ninety percent of evolutionary time had passed before life could emerge from water (or perhaps take water along) to populate the land. Until that epoch, life was to be found only below the water's surface, or near that boundary, or, at the driest surface locked within damp enclosing soil. Land life now has elaborate and specialized devices to avoid dryness. It remains true that the biopolymers themselves depend on water-like chemical bonds for their very existence. The hydrogen bonds in which a proton forms a positive electrostatic link between two negatively charged electron clouds, is the chemical bond of intermediate stability that lends the big polymers their subtle mix of stable and labile properties. If it were not water within which life grew it must all the same have been some other hydrogen-rich medium.

These logical inferences confirm our present findings: Life on Earth must have begun in or near water. That much seems sure.

SUGGESTION FOR FURTHER READING

Judson, H. F.: The Eighth Day of Creation. Simon and Schuster, N.Y., 1979.

Feb 1 /71
Beckenham

Down
Beckenham
Kent. S.E.

My dear Hooker

I return the pamphlets, which I have been very glad to read. — It will be a curious discovery if Mr Lowne's observation that boiling does not kill certain moulds is proved true; but then how on earth is the absence of all living things in Pasteur's experiments to be accounted for? — I am always delighted to see a word in favour of Pangenesis,

... died some day, I believe, will have a resurrection

Mr Dyer's paper strikes me as a very able Spencerian production. — It is often said that all the conditions for the first production of a living organism are now present, which could ever have been present. — But if (& oh what a big if) we could conceive in some warm

with all sorts of ammonia & phosphoric salts, — light, heat, electricity &c present, that a protein compound was chemically formed, ready to undergo still more complex changes, at the present day such matter would be instantly devoured, or absorbed, which would not have been the case before living creatures were formed. —

I enjoyed much the visit of your four gentlemen, i.e. after the Saturday night, when I thought I was quite done for. —

Your affec'
C Darwin

Henrietta makes hardly any progress, & God knows when she will be well. —

II. THE ORIGINS OF LIFE:
A BRIEF HISTORY OF THE SEARCH

THE EARLY QUESTIONERS

There is within modern science a curious anomaly, almost a paradox. Since the rise of modern science during the Renaissance, its fashioners have realized that the daily tasks of science could not be set by the great philosophical questions. The philosophers and the prophets over all history had raised the key questions: What is motion? What are being and becoming? What are the stars? Whence the Earth, life, man, and all the rest? But they could not supply growing insights, for all their keenness in setting great questions and opening logical conjecture. A more modest science could bring answers. But science by choice begins with small questions: How do bronze balls roll down inclines? What are the shapes of planetary orbits? What happens to the weight of charcoal as it burns? One had, above all, to supplement the inborn senses; the telescope, the microscope, the balance, the careful computations, these were the new tools of science.

With such tools and maturing skills, with concepts and analyses beyond the reach of the common language of the general philosopher, the scientists broke new ground in every direction. But they left aside the great questions, often the questions of origins and of ends, for such questions were not ripe for answers. It takes a mature discipline of geology, for example, to ask where mountains come

"But if (and oh, what a big if) we could conceive in some warm little pond, with all sorts of ammonia and phosphoric salts, light, heat, electricity, etc., present that a protein compound was chemically formed, ready to undergo still more complex changes, at the present day such matter would be instantly devoured or absorbed, which would not have been the case before living creatures were formed."
Charles Darwin's letter to Hooker, February 1871
(Copy courtesy of Prof. Melvin Calvin, University of California, Berkeley)

from; until the whole world is mapped, even the sea floor, and many individual histories have been teased out of the data, the worldwide picture of plate tectonics can hardly be drawn. Even now, the specialized sciences, with their diverse techniques and conceptual structures, do not often aim directly at the larger questions. There is a kind of disparity between the great end which science, as a whole, seeks — the full understanding of our universe and the human place within it — and the everyday or every decade problems which it can and does solve. The great questions are often put aside for another generation.

The problem of the origin of life is such a great question. Only in the last decades, since we have acquired a powerful molecular view of life's inner unity and a growing reach into space toward the old history of the planets and the Earth, can we begin to ask that question in so many words. The story of the growth of that topic within science, and the scale and scope of the community recently engaged with the issue are sketched here.

The idea of life arising from nonlife, the idea of spontaneous generation, had been commonplace for millennia. One had only to accept the evidence of the senses, thought the ancients: worms from mud, maggots from decaying meat, and mice from old linen. Aristotle had propounded the doctrine, along with Virgil and Lucretius. This teaching was accepted by a long line of western thinkers. Eastern ideas were similar. In the ancient Hindu scriptures life is described as having originated from nonliving matter. The *Rig Veda,* for example, pointed to the beginnings of life from the primary elements while the *Atharva Veda* postulated the oceans as the cradle of all living things.

The first pointed experimental investigation of the concept of spontaneous generation was carried out by Francesco Redi of Florence in 1668. His experiment was as simple as it was decisive. Once the jar of meat was covered with a veil of muslin, no flies could lay eggs on the decaying meat, and therefore it bred no maggots. All life is from the egg!

With the use of the microscope by Robert Hooke and Anthony Leeuwenhoek (ca. 1660–1700) a new false trail appeared. Many who used the new instrument saw many moving microorganisms grow amidst decaying vegetable matter but they were unable to explain their origin. The theory of spontaneous generation was thus kept alive for a century.

In 1860, the French Academy of Sciences offered a prize to anyone who would provide a decisive experimental result to halt the old controversy. Louis Pasteur's experiments of the 1860s with the swan-neck glass flasks are part of our scientific heritage (fig. II-1). Pasteur announced his results to the French in the following words: "Life is a germ, and a germ is life. Never will the doctrine of spontaneous generation recover from this mortal blow."

About the same time there came a clear insight from the field of organic chemistry. Perhaps it is premature to use that term, for in the mid-nineteenth century the chemistry of carbon compounds had not yet come of age. The great Berzelius in 1815 had argued that organic compounds were produced from the elements by laws differing from those governing the formation of inorganic molecules. According to him, organic compounds were produced under the influence of an essential vital force and therefore could not be produced artificially. But Wöhler's classic experiment of 1828, in which a product of animal metabolism, urea, was produced by the heating of ammonium cyanate, weakened the sharp distinction between the organic and

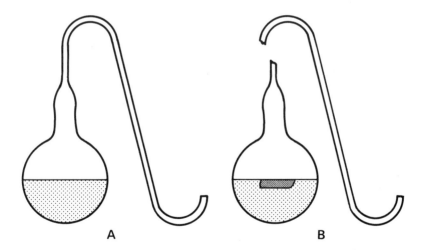

A   B

Figure II-1.– *In the 1860s, Louis Pasteur showed that life did not arise spontaneously. The intact swan-neck flask (a) remained sterile, while the one with the broken neck (b) did not.*

inorganic. Similarly, in 1845 Kolbe by a series of stepwise reactions produced the familiar acetic acid, surely a "genuine" organic compound, from carbon disulfide, which in turn had been prepared by reacting carbon with sulfur. The Chemical Abstracts today lists over 5 million organic compounds, elequent testimony to our unified understanding of organic chemistry, one single chemistry of carbon compounds.

THE EVOLUTIONARY SYNTHESIS

It appears that it was Charles Darwin who first formulated the modern approach to the origins of life, with a view of the circumstances, not of today, but of the distant past when the first life was somehow formed. He wrote in a private letter in 1871: "If we could conceive in some warm little pond, with all sorts of ammonia and phosphoric salts, light, heat, electricity, etc., present, that a protein compound was chemically formed ready to undergo still more complex changes, at the present day such matter would be instantly devoured or absorbed which would not have been the case before living creatures were formed." In short, the logical needs for the *origin* of life include the *absence* of life: a sterile environment was exactly what was present then and what is utterly unknown in the biosphere today. But for 50 years such large ideas lay dormant. They were ahead of the state of biology and geology. The question was too grand. Pasteur's wonderful declaration is true for our geological epoch; the ancient epoch when life originated, which is not at all the present natural life-filled environment, was not brought under study.

In 1924, a young Russian biochemist published a preliminary account of his ideas on the chemical origins of life. In a booklet entitled *Proiskhozhdenie Zhizny,* he pointed out that the complex combination of manifestations and properties so characteristic of life must have arisen in the process of the evolution of matter. A. Oparin had learned Mendeleev's ideas on the possible origin of hydrocarbons from the carbides in the crust of the Earth, and injected into his own thinking a new notion concerning the reducing nature of the early atmosphere. To Oparin's great credit, this observation was made before the astrophysicists had realized that the stars were 90% hydrogen.

In 1928, J. B. S. Haldane, the British biologist, independently of Oparin, wrote a classic paper, "The Origin of Life." Haldane speculated on the early conditions suitable for the emergence of life. According to him, when UV light acted upon a mixture of water, carbon dioxide, and ammonia, a variety of organic substances were made, including sugars and apparently some of the materials from which proteins are built up. Before the origin of life they must have accumulated until the primitive oceans reached the constituency of a hot, dilute soup. Haldane gave us the concept of the "primordial soup."

Almost 20 years after Haldane's publication, J. D. Bernal of the University of London conjectured before the British Physical Society in a famous lecture entitled *The Physical Basis of Life* that clay surfaces were involved in the origin of life. He was looking for ways and means by which the primordial molecules in the hot, dilute soup could be brought together to give rise to polymers capable of replication. A physicist and crystallographer by training, Bernal was particularly attracted to the role of surface phenomena in the origin of life. He argued that favorable conditions for concentration, which may have taken place on a very large scale, were provided by the adsorption of organic molecules on the fine clay deposits. The role of clay in primordial organic synthesis is today a lively area of investigation.

THE EXPERIMENTAL ERA: ORIGINS ENTER THE LABORATORY

In the early fifties, the Oparin-Haldane hypothesis, enriched by Bernal's ideas, was being widely discussed. Many experimenters were interested in putting it to the test. We must, however, recognize that over a long period of time innumerable experiments had been performed to generate organic molecules of biological interest. For example, Haber, in 1917, exposed a mixture of gases to electric discharges and postulated that any compound which could plausibly be synthesized would come out of such a system. (Rabinowitz, 1945, narrates many other examples, although they were not undertaken with the set purpose of studying the origin of life.)

The first reported experiment in the series of investigations to test the hypothesis of chemical evolution was done at Berkeley in

1950 by Calvin and his associates. The availability of radiocarbon 14 and the Berkeley cyclotron provided the ideal tools for such an experiment. A mixture of carbon dioxide, water, and hydrogen was exposed to 40 MeV helium ions from the 60-in. cyclotron at Berkeley. (Unfortunately no nitrogen was added to the mixture of gases irradiated.) A curious fact about this experiment was that although they referred to Oparin's idea of a reducing atmosphere, they used an oxidized source of carbon. Among the products identified were formaldehyde and formic acid.

Soon after the results of this experiment were reported in *Science* in 1951, there appeared a detailed paper by Harold Urey on the early chemical history of the Earth's atmosphere. In this paper Urey clearly defined the conditions under which a primitive atmosphere may have arisen and argued from the abundance of hydrogen in the universe, the rate of escape of the lighter elements, and equilibrium constants that the early atmosphere must indeed have been reducing. This paper was soon followed by the now-celebrated experiment by Urey's graduate student, Stanley Miller, in which methane, ammonia, and water were exposed to electric discharges (fig. II-2). Among the compounds formed and identified were four of the amino acids commonly found in protein: glycine, alanine, aspartic and glutamic acid.

Over the years, although new evidence on atmospheric outgassing and the geological conditions of the early Earth has led us to reconsider our thinking on the true nature of the Earth's primitive atmosphere, Urey and Miller had set the pattern for most of the synthetic experimental work in chemical evolution.

THE SPACE AGE

The scientific inquiry into the origin of life soon began to receive worldwide attention. In 1957, under the sponsorship of the International Union of Biochemistry and the leadership of Oparin, who was at that time its vice president, an international conference on the subject was convened at Moscow. At this meeting a number of biochemists were gathered who reported on their theoretical or

Figure II-2.– *An apparatus used to produce amino acids from methane, ammonia, and water by electric discharge.*

experimental investigations on the origin of life. The mammoth volume of 691 pages bears testimony to the extent of the work and the depth of interest of the international scientific community.

With the establishment of the U.S. space program in 1958, a dedicated effort in space biology began to emerge. In an authoritative document, the Space Science Board of the National Academy of Sciences declared that the search for extraterrestrial life was a prime goal of space biology. "It is not since Darwin and, before him, Copernicus, that science has had the opportunity for so great an impact on the understanding of man. The scientific question at stake in exobiology is the most exciting, challenging, and profound issue not only of the century but of the whole naturalistic movement that has characterized the history of Western thought for over 300 years. If there is life on Mars, and if we can demonstrate its independent origin, then we shall have a heartening answer to the question of improbability and uniqueness in the origin of life. Arising twice in a single planetary system, it must surely occur abundantly elsewhere in the staggering number of comparable planetary systems."

In 1963, the second conference on the Origin of Life was held at Wakulla Springs, Florida, under the sponsorship of the National Aeronautics and Space Administration. Much progress had been made since 1957. Many of the molecules of biological significance had been synthesized. The pathways for their chemical origin had been outlined, the conditions of reaction well defined, and the analytical techniques developed to a high degree of refinement.

By now, primarily with well-planned support from the Space Sciences Division of the National Aeronautics and Space Administration, several laboratories across the world were engaged in sophisticated experimental programs related to the origin of life and to life beyond the Earth. Papers on the subject appeared in journals as diverse in discipline as microbiology and astrophysics. The landing of a man on the Moon and the availability of lunar samples for analysis intensified the geochemical aspects of the program.

The year 1970 appeared a most appropriate time to organize the Third International Conference on the Origin of Life. This meeting was held at Pont-a-Mousson, France, about 250 km east of Paris. Over 150 researchers from several countries around the world were present. At this meeting was also born the International Society for the Study of the Origin of Life, with Alexander Oparin as its first president.

An accidental event of considerable importance at about this time was the fall of the Murchison meteorite in Australia. The techniques developed for the analysis of the lunar samples were soon applied to this fresh carbonaceous chondrite. Although meteorites had been analyzed before this fall, the finding of equal amounts of D- and L-amino acids and nonprotein amino acids in a meteorite provided unambiguous evidence of nonbiological and indigenous organic compounds for the first time. Prebiotic organic matter had been discovered in the solar system.

In 1973, the Fourth International Conference of the Origin of Life was held at Barcelona, Spain. The subject of chemical evolution had come of age. A quarterly journal, *Origins of Life,* devoted to the interdisciplinary approach to the subject, has been published since 1974. The *Journal of Molecular Evolution* and *Biosystems* often publish papers on the biochemical aspect of the question. *Precambrian Research* highlights those aspects related to the Archaen sediments.

The pace of research has quickened. The Viking Mission to Mars gave man the first opportunity to sample the soil of another planet and search for clues for life. The results of the Martian program were carefully analyzed at the Fifth International Conference on the Origin of Life which was held in Kyoto, Japan, in 1977, and, at the sixth meeting in Jerusalem in 1980.

Since that time, the growing interest in the subject has been demonstrated by the number of scientists from different disciplines and from different countries who have joined the Society for the Study of the Origin of Life. Apart from the international meetings held every 3 years, regional meetings in the United States, Europe, Japan, and India keep alive the expanding interest in the subject, of fundamental importance to all science.

SUGGESTIONS FOR FURTHER READING

Bernal, J. D.: The Origin of Life. World Publishing Co., Cleveland, Ohio, 1967.

Haldane, J. B. S.: The Rationalist Annual, 1929.

Oparin, A. I.: The Origin of Life on Earth, Third edition. Academic Press, New York, 1957.

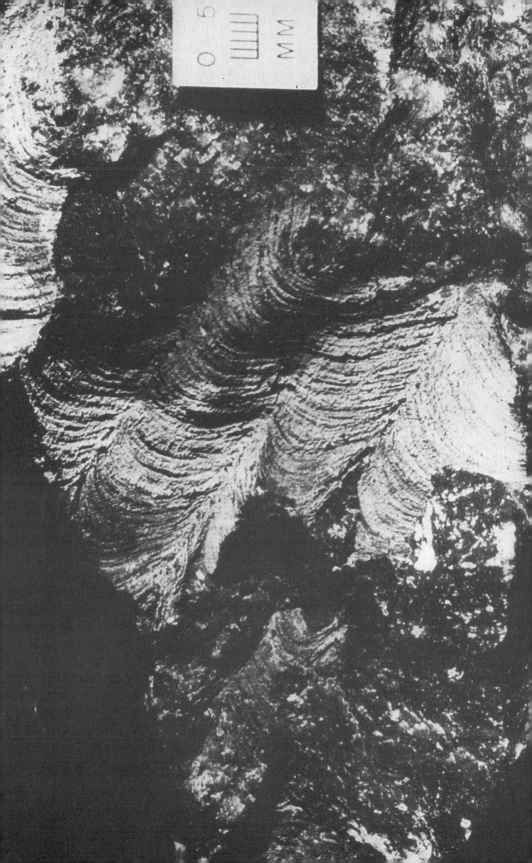

III. THE TWO RECORDS:
IN THE ROCKS AND IN THE CELLS

From the perspective of the long sweep of human history, over the past two centuries or so, less than a dozen human generations, scientists have begun to study, and to comprehend, the history of life on this planet. Progress has been impressive — nearly 200,000 species of fossil organisms have been discovered and described; the evolutionary continuum that links life of the modern world to that of earlier biotas has been extended far into the geologic past; and great strides have been taken toward deciphering the timing and nature of the major events in the development of life on Earth.

Early investigators interested in the history of life concentrated on those problems most readily amenable to investigation, and already by the mid-1800s when Darwin's *On the Origin of Species* first appeared, the broad outlines of the history of animals and plants were rapidly coming into focus. Indeed, it is this fossil record — that of the "Phanerozoic Eon" of Earth history, the ages of trilobites, coal swamp flora, dinosaurs, and the like — that normally comes to mind when one thinks of the history of life on Earth. Yet the Phanerozoic extends only a scant 600 m.y. into the geologic past (fig. III-1). But this is only 15% of geologic time. What came before, during the first 85%? How long before the advent of visibly large life did the earliest organisms first appear? And what does the geologic record reveal about the origins of life itself? These questions and others like them have long been pondered — indeed, Darwin regarded their solution as a necessary prerequisite to ultimate acceptance of his theory of evolution — but it has only been within the past quarter

Finely layered stromatolitic structures, of bacterial and/or blue-green algal origin, from Early Precambrian rocks near Bulawayo, Rhodesia, about 2600 million years in age. These structures are among the oldest stromatolites — and thus among the oldest fossil evidences of life — now known.
Photo courtesy of J. W. Schopf.

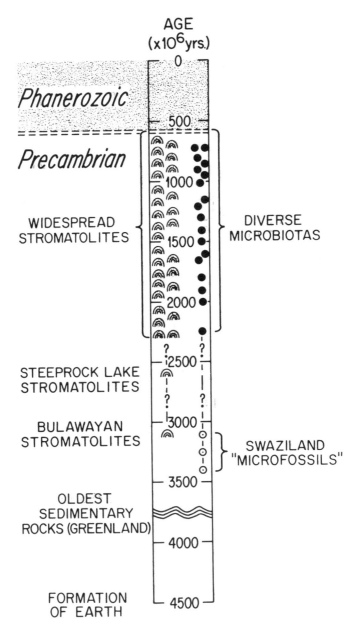

Figure III-1.— *A summary chart showing the known geological distribution of Precambrian fossil microorganisms (microfossils) and stromatolites. Photo courtesy J. W. Schopf.*

century or so that they have also proved amenable to scientific inquiry. Here, too, progress has been impressive — in recent years the known fossil record has been extended further and further into the remote past. The oldest known fossils are about 3.5 b.y. old, an age approaching, but markedly less than, the oldest rocks (approximately 3.8 b.y. old) on Earth, only one billion or so years younger than the age of the planet itself.

THE AGE OF THE EARTH

Knowledge of the age of the Earth is of fundamental importance to our understanding of the time when life originated. To fix the age of the Earth, one must determine the isotopic abundances of certain elements.

The isotopic composition of lead (Pb) at the time of formation of the solar system can be accurately determined by measurements of Pb in certain undifferentiated meteorites rich in water, carbon dioxide, and other volatile materials. Assuming the mantle source of terrestrial lead has always been a well-mixed reservoir containing uranium (U) and thorium (Th), decay of ^{238}U, ^{235}U, and ^{232}Th will enrich this primordial lead in the radiogenic daughter isotopes to the extent observed in modern samples of lead from Earth's mantle in a time of 4.43 b.y. Measurement of lead samples of various ages, extending back to 3.8 b.y. ago, shows that the assumption of a well-mixed reservoir is a good, but not perfect, approximation. Correction of the calculated age of the Earth for the observed deviation from a uniform source leads to an age about 100 m.y. older; i.e., 4.53 b.y.

Accurate measurements by U-Pb, and other isotopic clocks in meteorites, yield 4.55 ±0.02 b.y. for the time of formation (or differentiation) of their parent bodies. Measurement of the decay products of extinct radioactivies; e.g. aluminum (^{26}Al) and iodine (^{129}I), shows that those bodies in the solar system from which the meteorites were formed are within about 1–100 m.y. of the time at which the forming solar system itself was produced.

Formation of the Sun and planets post-dates this time. Thus, the age of the Earth is firmly bracketed between 4.65 b.y. and

4.43 b.y. The complete accumulation of the Earth is thought to have taken somewhat longer than that of the meteorite parent bodies, and it is therefore unlikely that the Earth is older than 4.55 b.y. Most likely the "corrected" lead isotope age of the Earth (4.53 b.y.), as discussed above, is within 50 m.y. of the time at which the Earth was formed and differentiated into silicate mantle and iron core. The real uncertainty in this figure could easily be 40 m.y., but is unlikely to be as large as 100 m.y. A value and range of 4.50 ±0.1 b.y. would seem conservative. This implies at least 0.5 b.y. — indeed, something like 0.6 or 0.7 b.y. — between (1) the formation of the Earth as a solid differentiated planet and (2) the oldest known rocks.

VISIBLE TRACES IN THE ROCKS

What then is known about the history of life on Earth? What were the trends that shaped the course of evolution, and over what time periods did they occur?

Two generalizations are clear. First, the history of life on Earth is on the whole the history of *microscopic,* rather than of visible, organisms. Because of our naturally anthropocentric myopia, as well as the relative ease with which fossils of the larger plants and animals can be discovered and studied, the history of organisms large enough to be seen with the unaided eye has received a disproportionately large share of scientific attention. The real situation is vastly different. Indeed, it is now known that the Earth's biota was composed solely of microscopic forms of life and their colonies for nearly 85% of the total history of life on this planet. All larger organisms are by comparison recent additions, interesting and significant, but they have been preceded by a long and well-developed evolution of microscopic forms.

Second, the type of life inhabiting the globe and the nature of evolutionary trends through time depend upon the peculiar geology of our planet. Earth, unlike all other known bodies of the solar system, is an aqueous planet, some 71% of its surface being covered by a thin, watery veneer. It is thus not surprising that liquid water (H_2O) is the major component of every known form of life. Water is the "universal solvent," the fundamental medium without which life as we know it could not occur. As a necessity for life, then, water serves also to limit life, and one of the principal themes that characterized the evolution of both plants and animals during Phanerozoic time was the development of structures and biochemical processes that enabled these life forms to spread to the land surface where water was in short supply. Indeed, many of the major evolutionary innovations of the Phanerozoic concerned the relations between water and life, such matters as the developments of lungs and of hardshelled eggs in animals, and the origin of seeds and of specialized pollenation mechanisms in plants, both once novel biological solutions to the scarcity of water on land.

Another important generalization concerning the fossil history of life is its unevenness. Some portions of the record are well documented and understood, but others are nearly unknown. It is a general, but not a perfect rule, that the older the material, the poorer the record. For markedly different phases of the fossil record, "eons" can be recognized:

1. The classical fossil phase: The Phanerozoic Eon, extending from about 600 m.y. ago to the present, is far better understood than any of the earlier phases; literally thousands of richly fossiliferous deposits of this age are known, units that collectively provide a sound and rather detailed basis for understanding the major aspects of the history of life.

2. Before the classical fossils: The Proterozoic Eon, extending from about 2500 m.y. ago to about 600 m.y. ago, when the Phanerozoic began, is understood only in outline; the total fossil record now known consists of three kinds of deposits: (a) There are about a dozen latest Proterozoic (690 to 600 m.y. old) fossiliferous deposits with large organisms primarily as sandstone impressions. (b) There are also about 150 microfossiliferous deposits (fig. III-2), spread somewhat unevenly throughout the eon (very roughly, 10–25 deposits per 100 m.y. during the later Proterozoic and only 1–5 deposits per 100 m.y. within the earlier portion of the eon). And (c) there are hundreds of limestones and dolomites that contain the structures called stromatolites. These structures are layered, commonly mound-shaped, sedimentary rocks. They were built over time through the growth and metabolic activities of whole communities of microscopic organisms; but, with rare exceptions, they do not contain fossil remnants of the bodies of the microorganisms responsible for

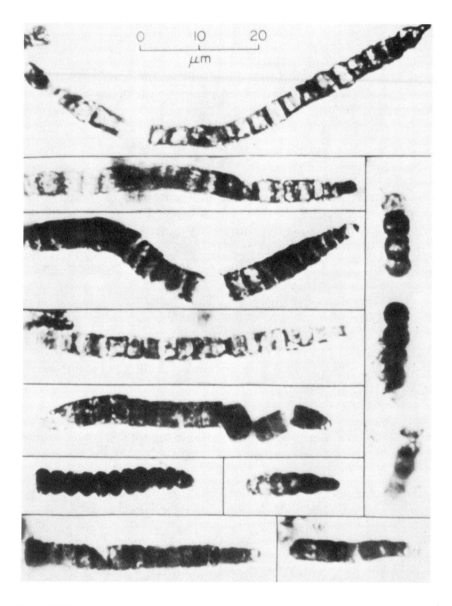

Figure III-2.– *Prokaryotic microfossils about 850 m.y. old (Late Precambrian) from carbonaceous black chert of the Bitter Springs Formation of central Australia.*

their construction. They are traces, rather than fossil organisms themselves. Research in this phase is vigorous and ongoing; over the past two decades it has resulted in substantial increases in our knowledge about Proterozoic evolution.

3. The ancient phase: The later portion of the Archean Eon, extending from 3.5 to 2.5 b.y. ago, is very poorly known. Here, the whole fossil record consists only of some seven or eight stromatolitic deposits, and of a few units known to contain microfossils or suggestive microfossil-like objects. Indeed, only one diverse, cellularly well-preserved microbiota has as yet been detected in rocks of this age. The oldest assured, bona-fide records of life now known are those contained in the sediments of the Warrawoona Group of Western Australia (fig. III-3). Those rocks are approximately 3.5 b.y.

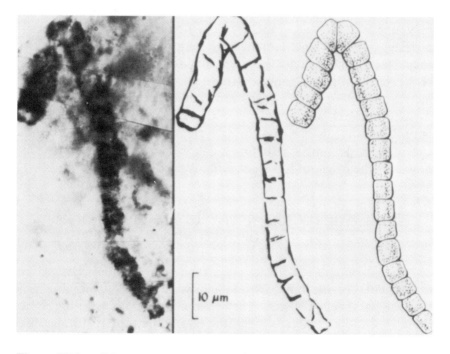

Figure III-3.– *Filamentous, prokaryotic (i.e., bacterium-like) microfossils in petrographic thin sections of carbonaceous chert from the Early Precambrian (ca. 3.5 b.y. old) Warrawoona Group in the North Pole Dome region of the Pilbara Block, northwestern Western Australia. Photo courtesy of J. W. Schopf.*

old; evidence indicates that diverse types of microorganisms, probably including photosynthesizers, were already extant at that time.

4. The most ancient phase: The Early Archean (Hadean), extending from the time of formation of the Earth, 4.5 b.y. ago, to about 3.5 b.y. ago, is all but unknown; only one rock unit has as yet been detected in this oldest portion of the geologic column, and that unit (the Isua Supracrustal deposits of western Greenland) has been severely altered by postdepositional motion, heat, and pressure. Indeed, the sequence has undergone at least five stages of alteration, including severe metamorphism (technically of amphibolite-grade, viz., to 500°–600° C, and 2–8 kilobars), a sequence of events so severe as to make meaningful interpretation of the Isua "fossil record" virtually impossible (if, in fact, organisms that could give rise to a fossil record actually existed in Isua time). Isua has been studied mainly over the last 5 years.

Thus, the fossil record as now known − rich and varied as it is in younger rocks, and scant to nearly nonexistent in older rocks − provides only the most limited insight into the timing and nature of those events that led to the origin of life. The evidence establishes that communities of complex, diverse microorganisms were extant at least as early as 3.5 b.y. ago, and that these organisms resembled modern bacteria in morphology, in ecology, and perhaps in biochemistry. Certainly, the origin of living systems must have occurred some time earlier, perhaps long before 3.5 b.y. ago. When it occurred, and how long the origin-of-life event took, is certainly not known.

THE CARBON CHEMISTRY OF ANCIENT ROCKS: AN OPPORTUNITY

As just discussed, rocks more than a few billion years old are scarce, not because only few of them were made but because their survival has been difficult. The examples that we have are, quite literally, battered veterans. Even the layered pattern of stromatolites rarely remains. If the rocks themselves have been mostly destroyed or significantly altered, it follows that their organic molecular constituents, which are much more fragile, must be in relatively poorer shape. Delicate information-rich biological macromolecules have *not* a chance of survival intact. On the other hand, not all organic matter has been so drastically altered that the study of organic geochemistry of ancient rocks is reduced to the study of plain graphite. There is a continuum between the extremes of "DNA" and "graphite." Sedimentary organic materials must inevitably move along this continuum as time progresses. As we move back more than 2.5 b.y. before the present into the Archean, the question arises as to what we will find first: The origin of life? Total degradation of all organic material to uninformative graphite? Or the end of the record? For at least 15 years, every generally accepted minimum date for the appearance of life on Earth has been based on morphological evidence alone (stromatolites, microfossils), i.e., visible or microscopic traces, not molecular ones. Chemical analyses have been viewed as suggestive, but not compelling. The approach suggested here acknowledges the preeminence of such morphological evidence when it comes to a yes or no question about the existence of life, but then works toward more insight. Specifically, if micropaleontologists can show where life existed, then perhaps organic geochemists can further support that interpretation *and* can determine *what kind* of life it might have been.

The most important advance in organic geochemistry during the past 10 years has been the emergence of a coherent view of the nature and role of the substance called kerogen. This material is a blackish and insoluble macromolecular complex dispersed in sedimentary rocks and comprising the great majority of organic matter in sediments. It has effectively resisted many efforts at detailed macromolecular structural analysis, presumably because it lacks the regular structure like coal or the asphaltene particles in petroleum. But it is now recognized as a product of geochemical reactions occurring over a very long period of time. "Protokerogen," an organic substance composed mainly of cellular degradation products, is formed today by living microbial communities in modern sediments. It is added to and modified by a series of reactions grading from the microbiological through the high-temperature geochemical. The present level of understanding is not complete and does not fully extend into the Precambrian, but it raises questions regarding chemical fossils and points the way to useful investigations of kerogens.

The times of the origin of photosynthesis, of respiration, of other biochemical processes of great ecological significance are not now known. Morphological evidence alone is unlikely to be decisive in fixing these dates, but it does seem possible that both chemical and isotopic investigations might be of substantial use. Kerogen analyses might follow an approach already highly successful in other fields of geochemistry and concentrate on determinations of stable isotopes of the elements present. It seems likely, for example, that any shift in the mode of primary production of polyatomic carbon molecular skeletons — chains and rings and the like — would have been accompanied by some shift in the ratio of light to heavy carbon isotopes within the reduced organic matter.

Whatever else they might do, kerogen analyses must deal with the fact that differences observed between specimens are as likely to be due to postdepositional effects as to differences in the original communities. The mineralogy, as it pertains to the origin and evolution of the rocks, and the structural geology of a given rock unit must be considered in addition to its organic geochemistry if this is to be done adequately. Such investigations are becoming more frequent, with many investigators realizing that the methods they choose should be designed *both* to decode the chemical message which might describe the original community *and* to assess the state of preservation of that message.

It seems appropriate in the study of microfossils with single structural morphology, such as most of those found in the Archean, to seek chemical-supporting data to establish their biological origin. Combined electron microprobe-scanning, electron microscope systems can now detect C, N, O, and P in micron-sized objects within reasonable limits of error. Micropaleontologists, using, e.g., microprobe techniques capable of detecting elements with atomic numbers as low as carbon, may be able to resolve details of the original organisms that left their remains.

Just the same, Precambrian paleobiology has now made notable advances. These include the certain great antiquity of now uncontested stromatolites, the wonderful Ediacaran fauna, the Bitter Springs, Gunflint, Transvaal, Belcher Island, and Fig Tree microfossils, and the oldest "North Pole" finds from W. Australia. Some of these deposits contain an abundance of well-preserved forms, e.g., Gunflint, Bitter Springs; others contain only relatively few forms, most of which are broken "debris," e.g., Transvaal. What is noteworthy is that these micropaleontological and organic geochemical findings can be related in some degree to living analogues.

The pursuit of kerogen investigation, together with related biological and geological studies, seems very likely to fulfill its promise. Organic geochemists now appear to have a good chance to make significant contributions, less to the blunt question "when did life arise?" than to the much more detailed set of questions dealing with the biochemical natures of the ancient communities once they are disclosed to the searching paleontological eye.

THE RECORD IN LIVING FORMS

While the paleontologist studies the fossils that the old organisms left in the rocks, the biochemist would like to study the biological processes in those earliest organisms. The descendants of those ancient organisms are alive today, and they, with their molecular traditions from the past, can be the subject of the biochemist's investigations. Some of these traces exist in a form much changed from the original, while others appear to have been handed down to generation after generation, practically unchanged. One important simple tenet of evolutionary theory is this: If several organisms are found to share a certain trait, that trait was most likely inherited from a common ancestor rather than evolved independently on several separate occasions. But certain important features are now shared in identical form by every organism extant. For instance, all life contains one set of amino acids, the building blocks of proteins, and one set of nucleotides, the building blocks of nucleic acids (ribonucleic acid (RNA) and deoryribonucleic acid (DNA), the genetic material). The main features of the complex system which serves to pass genetic information from one generation to the next are shared by all organisms. That is a clean example of what biochemistry can tell us about early common ancestors of all life: They contained nucleic acids, spelling out a genetic code for proteins, and they passed these down to all life forms present today. The random reorderings of amino acids and nucleotides observed in functionally related polymers must have evolved over the eons through various shuffling and mutational events. Locked in the arrangement of the monomers in various polymers in the cells of organisms is a coded record of the evolution of these organisms. The key to reading this record lies in modern methods for "sequencing" or determining the detailed arrangement, one after the other, of nucleotides in nucleic acids and the amino acids in proteins. Since the information for the sequence of amino acids in proteins is carried in the nucleic acid hereditary material, the sequences of both these polymer classes reflect evolutionary changes handed down from organism to organism. Sequencing the proteins or the nucleic acids allows the estimation of genealogical relationships among organisms presently alive. We can trace relationships among a group of organisms back to a common ancestor of that group that lived hundreds of millions or even billions of years ago.

The way this is done is to compare the sequences of related proteins or nucleic acids from a number of organisms. The degree of difference or diversity between these sequences is determined. The information concerning the degree of diversity among sequences can be used to estimate the order and perhaps the relative times of divergence of species from their ancestral relatives (fig. III-4). We can estimate how old a particular family of organisms is by determining how diverse the sequences within chosen molecules are among its member species. If a family exhibits a relatively high level of sequence diversity, it is held to have existed as a group for a relatively long time.

THE KINGDOMS OF LIFE

Genealogical relationships traditionally have been defined by such characteristics of organisms as shape, photosynthetic ability, and mode of cell reproduction. Now we use the biochemical record to define groups in terms of shared genetic information. For many years, all life was held to be divided into only two major kingdoms: the plant kingdom, studied by botanists, and the animal kingdom, studied by zoologists. More recently we recognized that all plant and animal cells exhibit fundamental properties not shared by bacterial cells. The present view groups all cellular life into two major divisions: the prokaryotes and the eukaryotes. Prokaryotic cells, all bacterial, are generally small, simple, relatively undifferentiated cells. Eukaryotic cells, which make up all the plants and animals, fungi,

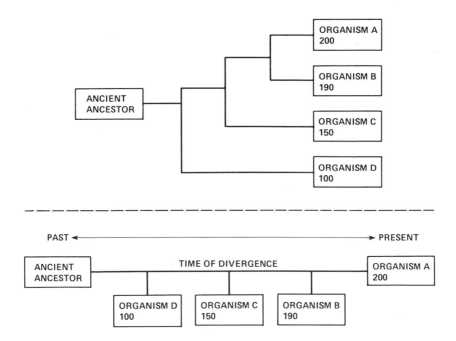

Figure III-4.— *Genealogical trees can be deduced from molecular-sequence information. Each box in the figure represents an organism, and the numbers indicate how many of the building blocks of a 200-block-long molecule in the organism are in the same arrangement as this molecule in organism A. Thus, the molecule in question in organism B has many building blocks (190) in an arrangement identical to organism A's molecule. This indicates they diverged from one another relatively recently. Organism D's molecules, on the other hand, has fewer blocks (100) in the same order as organism A's, indicating that they diverged from one another a long time ago. Organism C exhibits an intermediate time of divergence.*

and algae, are generally larger, more complex, and internally differentiated (containing several types of internal organelles which themselves are related to the free-living bacteria).

The Eukaryotes

One evolutionary puzzle that has responded to the biochemists' scrutiny is that of the origin of the eukaryotic cells (the nucleated cells of all larger forms of life). The currently accepted theory of the origin of these cells (the serial endosymbiotic theory) postulates that they arose by symbiotic association between some unknown pre-eukaryotic cells (urkaryote) and certain types of prokaryotes, then free-living (fig. III-5). These symbioses gave rise to structures recognizable today as intracellular organelles, introduced in one step, not incrementally evolved. The organelles of eukaryotes, thought to have originated from living bacteria, are mitochondria, which react with chemical substrates and atmospheric O_2 to produce chemical energy within all types of eukaryotic cells, and the plastids, subcellular sites of photosynthesis which convert radiant energy from light to chemical energy in plants and algae. All these quite complex organelles have their own nucleic acid genetic material, and they produce their own distinct proteins. The sequence of some protein molecules in organelles has been compared to that of similar proteins in extant prokaryotes. The amino-acid sequences of certain proteins in the plastids which have been examined indicate a close relationship to certain prokaryotes, particularly to the photosynthetic oxygen-producing cyanobacteria and prochlorons. The relationship of the

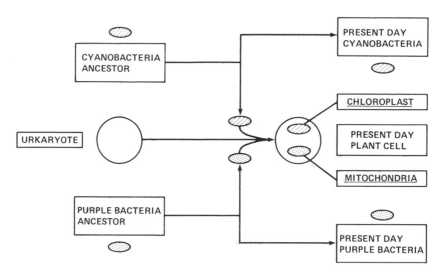

Figure III-5.— *A schematic illustration of the endosymbiotic theory of the origin of plant cells. Bacterial entities from the cyanobacteria (blue-green algae) group and the purple bacteria group were engulfed by an urkaryote and formed today's chloroplasts and mitochondria, respectively. Animal cell mitochondria probably have a similar origin, although their ancestry has not been clearly established.*

mitochondria to prokaryote cells is not as clear, though the mitochondria exhibit a relationship to the purple bacteria. As a whole, though, the information contained in the biochemical record to date strongly supports the endosymbiotic theory of the origin of eukaryotic cells. Eukaryotes can be seen as evolutionary mosaics, depending on contributions joined from several distinct lines of descent.

The Prokaryotes

Study of nucleotide sequences (16S RNA) in a large number of bacteria has lately revealed their enormous biochemical diversity. One group is called the eubacteria (true bacteria) and another called the archeabacteria (old bacteria). Still a third group, called the urkaryote, is represented by that part of the eukaryotic cell which is external to the organelles. The eubacteria and archeabacteria are no more closely related to each other than each is related to the third component, the cytoplasm. There are a great number of extant eubacterial species, but relatively few known archeabacterial species. Despite the paucity of extant species, the archeabacteria appear to be an ancient group which exhibit as much diversity as the eubacteria. Its members are generally restricted to unusual niches in the environment, hinting that the group might have originated during a period of Earth's history when prevailing conditions were different than they are today. The two bacterial groups together reveal greater biochemical diversity than all the extant eukaryotes.

The Antiquity of Biochemical Traits

The biochemical record can also yield inferences about the traits or phenotypic features possessed by very early organisms. A trait is an unlikely candidate for an ancient phenotype if the groups of organisms which now share the trait have all diverged from ancestral lines which did not possess it. A trait is probably not very old also if all extant organisms which possess it are closely related in terms of their biochemical record, indicating a recent time of divergence from ancestral lines. Conversely, if a trait is shared by groups so diverse that their biochemical record indicates a very ancient relative time of divergence, that trait may well have been possessed by very early organisms.

The first trait about which we have evidence concerns oxygen utilization by early organisms. There is some geological evidence indicating that on the very early Earth (Archean Eon) oxygen was much less abundant in the atmosphere. It is generally believed that significant quantities of oxygen first accumulated as the result of biological photosynthesis, a process in which free oxygen is a by-product. It is therefore thought that the first organisms must have been nonoxygen users, or anaerobes. The biochemical record supports this belief. Based on sequencing of nucleotides, anaerobic eubacteria are ancient compared with their aerobic counterparts. The major groups of the eubacteria are basically anaerobic, and the aerobic phenotype appears to have arisen relatively recently several times from various groups of anaerobic organisms.

Recent studies have shown that oxygen-using enzyme systems in eukaryotic organisms require orders of magnitude less oxygen to function than that in the present atmosphere. These findings suggest that aerobic biochemical processes may have arisen earlier than widespread aerobic life.

Even more fundamental is the nature of the energy source of the oldest organisms for which we have some clues. The biochemical record of the eubacteria has been examined to seek an origin for photosynthesis as a cellular trait, and suggests that photosynthesis is indeed ancient. Major eubacterial groups are photosynthetic; e.g., the purple photosynthetic bacteria, the cyanobacteria (blue-green algae), and green sulfur photosynthetic bacteria. Furthermore, nonphotosynthetic phenotypes have arisen several times from lines already photosynthetic (fig. III-6). For instance, the nonphotosynthetic, common, human intestinal bacteria, *E. coli,* most likely arose from the group of purple photosynthetic bacteria.

The usual view has been that the oldest forms of life were nourished not by internal biological photosynthesis but as heterotrophs that were supplied with energy by a rich environment, where organic nutrients that were abundant, were produced by processes which preceded life. We can find no sign of that early phase in the biochemical record among living cells. Photosynthesis in living cells goes back as far as the earliest groups of eubacteria. The lineages of the bacterial cells we know do not indicate whether heterotrophic or autotrophic life was first on the scene.

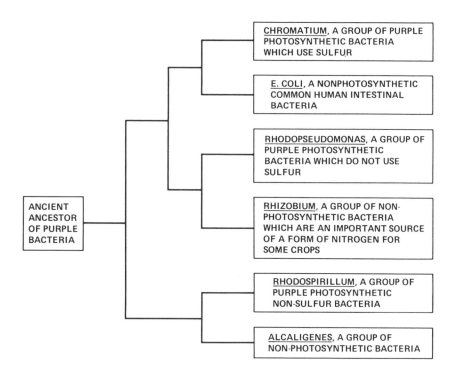

Figure III-6.– *Nonphotosynthetic organisms arose from photosynthetic organisms a number of times. This figure shows part of the genealogical tree of the purple photosynthetic bacteria drawn in the same way as the top of figure III-4. The bacteria known as E. coli, Rhizobium and Alcaligenes are all nonphotosynthetic, yet they appear to have independently arisen from the purple photosynthetic bacteria line of descent. This figure also illustrates the finding that aerobic lines appear to have arisen from anaerobic ancestry. E. coli, Rizobium, and Alcaligenes are aerobic, while the rest of the group is anaerobic.*

Finally we ask about that eukaryote component which is outside of all the cytoplasmic organelles – the nucleocytoplasm. It is quite distinct in its sequencing from both bacterial lines. That allows the idea that this urkaryote is the most ancient ancestral form we see, though it does not require that. We expect to find in that form only the features which the eubacteria, archeabacteria, and urkaryote share in common, and not much more. Those features present include of course the amino acids and their polymers as a class, the proteins; and the nucleotides and their polymers, the nucleic acids.

Apart from those generalities, the details of all three groups differ. The proteins, especially enzymes and their consequent metabolic pathways, the cell walls, and most of the membranes differ. This suggests that the three great groups diverged at the earliest stage, with not much but the building blocks, the genetic code, and its ribosomal machinery in common.

More than that we have not been able to read from the details of the similarities of living forms. Perhaps we should have expected no more. For all the forms we know now are cellular, compact; the earlier forms we are looking for have still to depend for some necessity upon features of the nonliving environment, whether it be for a source of chemical free energy, the means of replication and variation, or the simple preservation of high concentrations of key constituents. All the primitive cells, we infer, are microbial, enclosed, and can replicate using the full normal apparatus of bacteria; they derive free energy either from some part of incident sunlight or from an organic-rich world or both. There is still a big gap to explore. For that we must turn to the laboratory.

SUGGESTIONS FOR FURTHER READING

L. Margulis, Symbiosis in Cell Evolution, W. H. Freeman Co., San Francisco, 1981.

J. W. Schopf, editor, The Earth's Earliest Biosphere, Princeton University Press, 1983.

C. W. Woese, Sci. Amer., "Archaebacteria," vol. 244, no. 6, 1981, pp. 98–121.

IV. THE NATURAL EVIDENCE

In chapter III we have seen how the rock record provides information about life on the early Earth. And we have seen that this record leaves us a gap in time from about 3.8 b.y. ago to the time of Earth's origin at 4.5 to 4.6 b.y. ago. We have also discussed in chapter I the essential elements of life. But what were the conditions under which life originated? What was the early Earth like? To understand this we seek additional knowledge: we must know how the solar system formed and how the Earth formed from it. In addition, we must know the chemical behavior of carbon and its compounds in the prebiological environments. Only then will we have the knowledge necessary to bridge the gap and understand how life came to be.

THE BIRTH OF THE SOLAR SYSTEM

Before there were any stars there was only gas, and this gas was essentially just a mixture of hydrogen and helium. But by the time the Sun appeared, several previous generations of stars had added other heavier elements to this interstellar gas as a result of the synthesis of elements in stellar interiors, and in the catastrophic explosions of supernovae. We find evidence for this change in composition by examining the various types of stars in the galaxy. The oldest stars we find today (having formed at an earlier epoch) contain a smaller amount of heavy elements than does the Sun. The age of the galaxy is estimated to be close to 12 b.y., while the oldest stars with heavy-element concentrations similar to the Sun's have ages of only 6 to 7 b.y. The Sun itself is about 4.6 b.y. old.

Today we can examine environments where stars form and then try to reconstruct the conditions that preceded the appearance of the Sun and its retinue of planets. We find these environments among the

The solar magnetic field (white spiral lines) as it might have appeared 4.5 b.y. ago when created by the high spin rate of the Sun. This would have produced melting of rocks in the Asteroid Belt and elsewhere in the Solar System.

clouds of gas and dust in the spaces between the stars: the interstellar medium. As we explore these clouds, we must make some allowances for the further changes that have occurred since the Sun was born; however, most of what we find today appears characteristic of conditions that must have existed 4.6 b.y. ago.

This exploration, still in its infancy, is yielding a continuous stream of new information about the chemical composition, mass, and distribution of these clouds, and about the relationship of different types of interstellar clouds to the process of star formation. It is already clear that the sheer mass of the low-density interstellar material dictates that most of the chemistry that goes on in the Universe takes place in interstellar clouds.

If the material in the clouds were spread uniformly over all space, the concentration of matter would amount to about three hydrogen atoms per cubic centimeter. We consider two basic types of clouds: (1) diffuse clouds, which contain little dust, and where concentrations of gas molecules are very low and single hydrogen atoms are the dominant species; and (2) dark, dense clouds, which contain abundant dust, and where molecular hydrogen gas (H_2) is the dominant species. In the latter clouds the gas concentrations range from about one thousand to about ten million molecules per cubic centimeter. Presumably, the material in dense interstellar clouds is utilized in star formation. Relatively small stars, such as our Sun, are probably able to form almost anywhere in these massive interstellar clouds. The dust in interstellar clouds is not well characterized; there is evidence to suggest that it is composed of ice, silicates, graphite, and both simple and complex carbon-containing compounds. While there is relatively little information about the dust, there is a growing body of data about the molecules present in these clouds.

How are interstellar molecules produced? First, it is necessary to consider the environment in which synthesis might occur. The temperatures are very low, from $-236°$ C to $-173°$ C. In addition, the extremely low concentrations of molecules in the gas phase means that collisions (and therefore chemical reactions) are generally restricted to those that involve only two species (that are binary). These and other constraints have led to a model for the synthesis of interstellar molecules in dense clouds in which reactions are initiated by collisions of ubiquitous, high-energy cosmic rays with hydrogen and helium. These encounters produce reactive, positively charged species which initiate chains of reactions leading to formation of the observed molecules. In this manner complex molecules and, perhaps even interstellar grains, are constructed within the clouds. A list of molecules observed in these interstellar clouds is given in table IV-1. At this writing, more than 50 species have been identified. Glycine, since it is the simplest amino acid (the building blocks of protein), is of obvious interest for the origin of life. To date, the search for it in the interstellar clouds has been unsuccessful. Finally, in addition to the species listed in table IV-1, many other as yet unidentified molecules appear to be present.

Alternative schemes for synthesis exist. One of those is the suggestion that the molecules are not formed in interstellar clouds but rather are formed under relatively high temperatures and high-density conditions, such as in primordial solar nebulae. Those nebulae are clouds of more concentrated dust and gas that form from interstellar clouds and directly spawn suns. Regardless of the model chosen, examination of the compounds in table IV-1 leads to two important observations. First, the compounds are chemically diverse and structurally complex. Second, many of them are also known to be produced by abiotic synthesis experiments in the laboratory (see chap. V).

Clearly, the interstellar environment, as exotic and as seemingly inimical to chemical reactions as it may appear at first consideration, exhibits a rich chemistry which manifests itself in the production of organic compounds that, for the most part, are familiar to the

TABLE IV-1.– SPECIES IN THE INTERSTELLAR MEDIUM

Number of atoms in the species								
2	3	4	5	6	7	8	9	11
H_2	HCN	H_2CO	CH_2NH	CH_3OH	CH_3CHO	$CHOOCH_3$	CH_3CH_2OH	$HC_2C_2C_2CN$
CH	H_2O	HNCO	HC_2CN	NH_2CHO	CH_2CHCN		$(CH_3)_2O$	
CH^+	H_2S	H_2CH	CHOOH	CH_3CH	CH_3NH_2		CH_3CH_2CN	
OH	OCS	NH_3	NH_2CN		HC_2CH_3		$HC_2C_2C_2CN$	
CN	HCO	C_3N	CH_4		HC_2C_2CN			
CO	SO_2		C_4H					
CS	HCO^+							
SiO	HN_2^+							
SiS	C_2H							
NS	HNC							
C_2								

earthly experience. Although many scientists believe there is little basis for the speculation that individual interstellar organic molecules found their way, intact and unchanged, to the prebiotic Earth's surface, there is growing interest in the possibility that interstellar molecules and dust might be preserved intact in comets and, in altered form, in the carbonaceous meteorites.[1] These possibilities stem from the probability that all matter in the solar system originated from interstellar dust and molecules. To the extent that comets and carbonaceous meteorites contributed mass to the early Earth, interstellar organic compounds could have survived to take part in subsequent chemical evolution.

Just as biological evolution implies that all organisms on Earth have a common ancestry, so chemical evolution implies that all matter in the solar system had a common origin. Observations of various stages of star formation and evolution in dense interstellar clouds support the view that our Sun and solar system formed from interstellar dust and gas. Serious scientific consideration is being given to the following scenario:

> An interstellar cloud of dust and gas underwent gravitational collapse, perhaps triggered by a shock wave generated by a nearby supernova; thus began the chemical evolution of the nascent solar system. Cloud contraction led to the formation of the primordial solar nebula, an enormous spinning disc of dust and gas with the proto- (or newly forming) Sun at the center. Heating, associated with gravitational contraction, produced a thermal gradient in the nebula, possibly with temperatures in excess of $1300°$ C, close to the protosun. According to the model, the temperatures were about $300°$ C at the present distance of the Earth to the Sun, but remained low ($-150°$ C) at Jupiter's distance and beyond. While uncertainties exist about the temperatures in the nebula, the qualitative trend of decreasing temperature with increasing

---

[1] Carbonaceous meteorites are so named because of their high concentration of carbon (up to 6%). The nature of this carbon will be discussed in detail shortly.

distance from the protosun seems to be acceptable. Cooling of the hot gas in the inner nebula close to the protosun led to condensation of solid mineral grains. As interstellar gas and dust were drawn into the solar nebula they would have been heated to varying degrees, depending on their distance from the protosun.

Further condensation accompanied by aggregation of fine-grained material yielded planetesimals ranging in size from kilometers to tens of kilometers. Continued growth of these objects by accretion led ultimately to the formation of the solid bodies of the solar system. Formation of organic matter from gases containing carbon, nitrogen, hydrogen, oxygen, and sulfur could have accompanied the condensation processes; mineral grains could have provided surfaces to catalyze the synthesis. As previously noted, in the outer regions of the nebula, temperatures would have remained low ($\leqslant$-150° C). There, low-temperature accretion of organic and inorganic material into planetesimals could have taken place, allowing the preservation of the ices and other volatile compounds originally in the interstellar medium as well as the rare gases helium, argon, neon, krypton, and xenon. Thus, from the solar nebula came the Sun, the planets and their satellites, comets, meteorites, and asteroids. In general, it is thought that material accreted in the inner solar system originated at relatively high temperatures and was depleted in volatile substances, while the materials that accreted in the outer solar system were volatile-rich. The validity of this simple model has been and continues to be subjected to critical assessment; inevitably, as theory, experiments, and observations progress the model will undergo changes, perhaps so many that a new model will emerge. In the meantime, it provides a useful framework within which to discuss various aspects of inorganic and organic chemical evolution.

COMETS AND METEORITES

Comets occupy an especially interesting place within this framework. They may have been a source of part of the atmospheres of the terrestrial planets, and they are believed to have been the planetesimal-like building blocks for some of the outer planets and their satellites. Present knowledge places the origin of comets in the outer regions of the primitive solar nebula, both in and beyond the space now traversed by the giant planets (Jupiter, Saturn, Uranus, and Neptune). Perturbation of their original orbits, by the formation of these giant planets, is believed to have sent some protocometary bodies into the inner solar system (to collide with the Sun and inner planets — Mercury, Venus, Earth, and Mars) and others into orbits extending great distances from the Sun (up to 50,000 astronomical units (AU); 1 AU = 150×10^6 km, or the distance from the Sun to the Earth).

Comets consist of a nucleus, a coma, and a tail (see fig. IV-1). According to a current model, comet nuclei contain simple and complex organic molecules, and meteorite-like dust and rock imbedded in a matrix of frozen water, possibly solid carbon dioxide and other ices. As comets approach the Sun, heating occurs and the ices vaporize, ejecting volatile "parent" compounds (possibly water, carbon dioxide, methane, acetylene, ammonia, hydrogen cyanide, etc.) and entraining nonvolatile dust and rock from the nucleus. In the coma that results, interactions of the gaseous parent compounds with solar radiation can lead to physical and chemical processes that cause the partial to complete breakdown of the so-called parent molecules to "daughter products." The uncharged daughter products are observed in the coma, whereas the positively charged ones are observed in the tail. According to an alternative view, all the observed daughter products already existed "frozen" in the nucleus of the comet, and were simply released directly into the coma by evaporation. In addition to the species indicated in table IV-2, metallic elements (iron, silicon, magnesium, calcium, nickel, sodium, chromium) have been detected by means of spectroscopic analysis of comets that pass very close to the Sun, and of meteor showers associated with comets. The relative abundances of these elements suggest similarities between the chemical compositions of cometary dust and carbonaceous meteorites.

The nucleus of a comet is thought to be small, typically 1 to 10 km diam, but no direct measurement of a nucleus has ever been made. Nuclei appear as small points of light imbedded within the bright and extensive coma of the comet. The mass of nuclei could range from 10^{15} to 10^{18} g. The light from the visible coma and tail is emitted by atoms and molecules that have interacted with solar

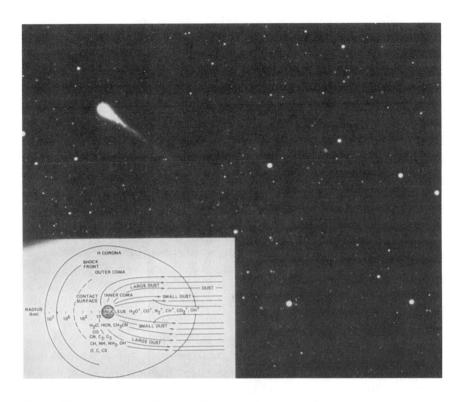

Figure VI-1.— *A comet showing the major features and species observed in the coma and tail.*

TABLE IV-2.— SPECIES IN COMETS

Coma			Tail	
HCN	CN	NH_2	H_2O^+	N_2^+
CH_3CN	CH	C_3	CO_2^+	CO^+
H_2O	OH			CH^+
$CO_2{}^a$	CO			OH^+
NH_3	NH			
	C_2			
	CS			

^{a}Suggested parent molecules, not detected.

radiation. The size of the coma is remarkable as measured by the light emitted by the atomic hydrogen in it. The coma may be $>1\times10^6$ km in radius. The tail, composed of dust grains and charged molecules, is even larger, $>100\times10^6$ km in length in some cases. When comets become visible in the inner solar system they can be, spatially, the largest objects in the sky, bigger than the Sun itself.

As mentioned above, comets are believed to be material condensed and accreted in the outer regions of the primitive solar nebula. Thus, a relationship may exist between interstellar matter and the dust and molecules that make up comets. If one compares the molecules observed in interstellar clouds (table IV-1) and in the coma and tails of comets (table IV-2) there do seem to be similarities between the populations. For example, both contain cyanide, and derivatives with a "CN" group. It is also possible that comets are related to some of the carbonaceous meteorites in that the latter objects, less rich than comets in various forms of the volatile elements and organic matter, may be derived from remnants of volatile-depleted, moribund comets. It is appropriate to note that if comets *do not* contain relatively unaltered interstellar matter, and if they formed at the outer edge of the solar nebula, where temperatures were sufficiently low to condense gases like carbon dioxide and water, then the presence of parent organic molecules in comets is difficult to understand. No widely accepted model exists for the chemical reactions that could have occurred in the solar nebula to yield the chemistry of comets. Indeed, in the absence of direct observations of the nucleus, our knowledge of comet chemistry is unfortunately sparse and model-dependent. Since comets may represent a chemical evolutionary link between the primitive solar nebula and the interstellar medium but are poorly understood, their direct study by space probes constitutes a high-priority objective for many scientists.

Unlike comets which have only been observed from afar, meteorites are rock samples of extraterrestrial origin that have survived passage through the atmosphere to the Earth's surface and are available for direct examination. Preserved in these objects are chemical, mineralogical, and structural information about the nature of the environments and the processes involved in their formation. Indeed, recent discoveries of anomalies in the isotopic composition of some elements (e.g., oxygen, aluminum, magnesium, noble gases) in meteorites even provide a connection with nucleosynthetic events that preceded the solar system, perhaps that triggered its formation. With few exceptions, the ages of meteorites fall within the range of 4.6 ±0.1 b.y. Since these objects constitute the oldest datable material now available, their study provides clues to the very early history of chemical evolution in the solar system. Although some uncertainty remains in identifying the source(s) of meteorites, there appears to be agreement that most were derived from asteroidal parent bodies, either in the main asteroid belt or those with Earth-crossing orbits or both. Some meteorites may be fragments of the inactive cores of ice-depleted short-period comets. According to a current scenario for solar-system origin, meteorites and the bodies from which they were derived (parent bodies) were formed as a result of the condensation and early evolution of planetesimals from the primordial solar nebula; these represent the building blocks from which solid planets and moons were assembled.

Because of turbulence and thermal and pressure gradients in the nebula, solid material that condensed at widely different radial distances and, therefore, different physical and chemical environments, could have been brought together and assembled into a common body. In this context, protoasteroidal and protocometary bodies may be viewed as components of a distribution of planetesimals that accumulated increasing proportions of ice and other volatile-rich phases. The accumulation of the diverse ingredients into parent bodies, possibly resembling asteroids, would have been accompanied by various processes which would have further influenced the chemistry, mineralogy, and structural features of the material and, to varying degrees, masked the features that would have been characteristic of primary solar nebula condensates and of originally interstellar material. Presumably, perturbation of a parent body, perhaps by collision with another object, yielded fragments of the bodies, some of which eventually fell under the influence of the Earth's gravitational field.

Meteorites can be placed in two general categories: (1) partially to fully differentiated and (2) undifferentiated objects. Differentiated meteorites exhibit strong chemical fractionation relative to average solar-system composition as represented by the Sun; they show clear evidence of having been derived from parent bodies that have undergone processes analogous to planetary core formation and volcanism. Evidently, partial melting of primitive undifferentiated material in asteroidal-sized bodies gave rise to the oldest basalts in the solar system at about 4.5 b.y. ago. Elucidation of the circumstances of early differentiation of some meteorite parent bodies and the nature of the heat sources involved may have much to tell us about the course of the Earth's differentiation to form the mantle and core, and its subsequent thermal history.

Undifferentiated meteorites have elemental abundances similar to those found in the Sun. Among these meteorites, the carbonaceous chondrites are closest to the Sun in bulk elemental composition and are considered to be among the least fractionated and, therefore, most primitive solid objects available for study in the solar system (fig. IV-2). It is noteworthy, however, that relative to the Sun, even the carbonaceous meteorites are depleted in hydrogen, carbon, nitrogen, and noble gases. Observations indicating less depletion of these elements in comets signify that comets are even more primitive bodies than meteorites.

Carbonaceous meteorites consist of complex assemblages of relatively fine-grained mineral and organic matter that reflect a broad range of elemental compositions and textures. This is indicative of wide variations both in the environments of origin for the various components and in the evolution of the respective parent bodies. For present purposes, we consider the classification of carbonaceous meteorites into three types: CI, CII, and CIII. Major differences among these types lie in their content of volatile elements and minerals of high-temperature origin; these are inversely correlated. Accordingly, the amount of organic matter increases in the order CIII, CII, CI, with the CIII containing about 0.5% and the CI having about 5% by weight. Similarly, minerals exhibiting a high-temperature history occur most abundantly in CIII meteorites, along with metals (iron and nickel). These minerals exist only in low to trace amounts in CII meteorites and are virtually absent in the CI meteorites.

Mixtures of clay-like minerals comprise the predominent minerals in CI and CII meteorites (50% to 80%) and a minor proportion in some CIII meteorites. These minerals resemble terrestrial clays in crystallographic structures, and the mixtures exhibit bulk elemental compositions remarkably similar to the pattern of solar abundances. Recent research suggests that like terrestrial clays, the clay-like materials in carbonaceous meteorites were formed in an aqueous

Figure IV-2.– *The Murchison carbonaceous chondrite. Most of our understanding of the organic matter in meteorites has been derived from studies of the Murchison meteorite.*

environment. Thus, the oldest known clays in the solar system were probably produced on parent bodies of carbonaceous meteorites.

The organic matter in carbonaceous meteorites occurs in various forms. A high-molecular-weight complex material characterized by insolubility in solvents and acids makes up the major carbon-containing component in all three types of meteorites. (Terrestrial sediments contain a material called "kerogen," which has similar characteristics but is obviously of different origin.) The source(s) and production mechanism(s) for this insoluble material are unclear, but may involve interstellar environments as well as environments in the solar nebula and on the parent bodies themselves.

Solvent-extractable organic matter in CII meteorites (i.e., the Murchison meteorite) is distributed among a variety of compound classes: alcohols, aldehydes, amines, amino acids, carboxylic acids, hydrocarbons (aromatic and aliphatic), ketones, purines, pyrimidines, etc. Carbon species found in the Murchison meteorite include the following:

1. A carbonaceous phase not affected by solvent
2. Carbonate
3. Hydrocarbons (aliphatic and aromatic)
4. Carboxylic acids
5. Amino acids
6. Ketones and aldehydes
7. Urea and amides
8. Alcohols
9. Amines
10. Nitrogen-containing heterocycles

More detailed information on these chemical compounds will be presented in the next chapter. At most, these compounds constitute 30% of the total carbon, and less than 0.5% of the total mass of the meteorite samples in which they are found. Some of these compounds have been sought and found in other carbonaceous meteorites, but only the Murchison meteorite has been studied in great detail because of the availability of samples and its relative freedom from terrestrial contamination.

The variety of types of compounds found and their molecular structures point to origins in nonbiological processes. However, the nature of the processes and where they occurred remain to be clearly established. Electric discharges and other gas-phase processes and gas-solid reactions requiring catalytic grain surfaces could have taken place both in the nebula and on parent bodies. An interstellar origin for some of these compounds should also be considered.

Recent isotopic studies of organic matter in carbonaceous meteorites have revealed that large differences exist in the deuterium/hydrogen, carbon-13/carbon-12, and nitrogen-15/nitrogen-14 ratios associated with *different* organic components within the same meteorite. Although the full implications of these findings remain to be elucidated, the magnitudes of the isotopic variations and their occurrences among different components strongly suggest that more than one source region and/or more than one production mechanism must have been involved.

Evidence that the clays in carbonaceous meteorites were produced in secondary aqueous alteration processes raises the possibility that some of the organic matter might also have been produced at the same time by alteration of preexisting compounds by water. Possibly, simple species (such as cyanide compounds), which occur abundantly in the interstellar clouds (table IV-1) and have been observed in comets (table IV-2), could have been present and served as precursors for some of the more complex molecules found in these meteorites.

From studies of interstellar dust and gas, comets, and meteorites the initial conditions in the solar nebula and its subsequent chemical evolution are being elucidated. Continuing investigations into the cosmochemical origins of organic matter are crucial because organic chemistry occurs throughout the cosmos, and the organic matter that results constitutes a molecular and isotopic record of the materials and processes involved in its formation.

We have outlined a scenario by which the building blocks of solar-system bodies may have developed from dusty and gaseous starting materials. We might ask if there are other solar systems in the universe. We feel that this possibility exists because of the frequent occurrence of binary and multiple stars (which now seem to be well over 50% of the star population), and the fact that the separations between most of these binary stars are comparable to the dimensions of our solar system. The inference is that stellar condensation tends to form more than one object; when the residual matter is insufficient to form a second star, planets may occur instead. We see this same tendency for multiplicity within our solar system. Only Venus and Mercury are without satellites, and this lack may be attributed to gravitational perturbations caused by the close proximity of these bodies to the Sun.

COMPARATIVE PLANETOLOGY

It is instructive to examine the Jupiter satellite system in this context. The Voyager spacecrafts have confirmed and extended the impressions gained from ground-based and Pioneer spacecraft observations. The four large satellites of Jupiter exhibit gradients in average density and other properties that are strikingly reminiscent of the gradient observed among the planets. The inner satellites, Io and Europa, have densities of 3.5 and 3.0, respectively, indicating a predominantly rocky composition. In contrast, the more distant satellites Ganymede and Callisto have densities of 1.9 and 1.8, respectively, suggesting compositions that include a high percentage of carbon, nitrogen, and oxygen compounds and, presumably, are dominated by water (ice).

The surface appearances of these objects substantiate this interpretation. The icy crusts of Ganymede and Callisto have evidently been unable to support the topography associated with large-impact craters, although crater densities in the smaller size range are close to saturation. The surface of Europa appears to be covered with a layer of water ice that has obliterated any trace of its history of early bombardment. The few craters that do appear are comparable in the number per unit area to that found on Earth. Io is so wreaked by continuous volcanism that its entire surface must be reworked on a time scale that is very short compared with 4.6 b.y. As a result of its tidally induced volcanic activity, this satellite appears to be extensively degassed, with the volcanoes possibly relying on sulfur dioxide as a working fluid instead of water.

While obviously much smaller than the Sun, Jupiter is apparently large enough to have caused the same gradient in the properties of its retinue of satellites as the Sun has caused in the planets (fig. IV-3). This fractionation of material can be attributed to the heat released during the formation of the central body. The resulting similarity between these two systems (Jupiter's and the Sun's) strengthens our intuitive feeling that the gross characteristics of the solar system are probably representative of those found elsewhere; with small dense inner planets possessing secondary degassed atomospheres. The exploration of the Saturn system by Voyager has introduced an important caveat: For this gradient in properties to exist, the central body must be massive enough to heat the surrounding space during the time the system forms, and the planets or satellites must be large enough to represent a homogeneous sample of the accreting material. Neither of these conditions was met in the case of Saturn (fig. IV-4), with the result

Figure IV-3.— *Jupiter, the largest planet in our solar system, has an atmosphere dominated by hydrogen, and has a retinue of satellites analogous to the planets of the solar system.*

that its moons are very different from those of Jupiter. By analogy, one might expect the planets of a red dwarf star to differ considerably from the planets in our own system. But even given another star like the Sun, what are the chances of finding another planet like

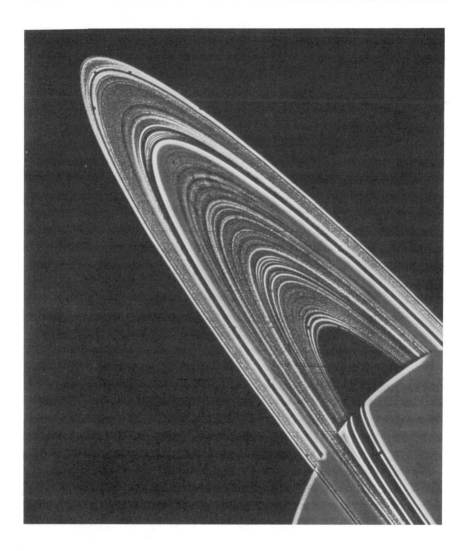

Figure IV-4.– *The ring system of Saturn is composed of ice.*

Earth? We are apprehensive about this question since our planet is so different from its neighbors.

The planets in our solar system that are most similar to Earth are Mars and Venus. Historically, this apparent similarity in gross properties led many scientists to anticipate that these planets might also be populated with some form of life. Yet, as we learned more about these two planets, this possibility became increasingly remote. Our considerations of the origin of life on Earth must therefore include some discussion of the unique properties of Earth itself. Why is our planet so different from its nearest neighbors?

We can quickly identify two basic characteristics that will determine most of a planet's development — size and distance from the Sun. If a planet is too large, hydrogen will not be able to escape from its gravitational field, and the result will be an object like Jupiter or Saturn, with a huge, dense atmosphere and no solid surface. At the other extreme, a body that is too small will not be able to retain any kind of atmosphere over geological periods of time. Examples of this end of the spectrum are Mercury and the Moon. Distance from the Sun becomes a part of this constraint, however, since a small body at a large distance could be cold enough to have an atmosphere. What this means is that the thermal velocities of gas molecules will be smaller than the velocity needed to escape from the body's gravitational field (given that the molecular weight of the gas is high enough and that the gas does not condense at the low temperature corresponding to this distance from the Sun). In our solar system, the best example of this situation is Titan, a satellite of Saturn, that has a dense atmosphere of nitrogen with a small amount of methane and traces of other compounds.

The inner planets — Mercury, Venus, Earth, Mars — have clearly been drastically affected by their proximity to the Sun. They are all grossly deficient in the volatile elements, having formed in an environment that was evidently at too high a temperature to permit the common compounds of these elements to condense. In contrast, the outer planets — Jupiter, Saturn, Uranus, Neptune — have retained large amounts of hydrogen and helium. Jupiter may even represent a cosmic mixture of the elements; i.e., the composition of this planet may be identical to the composition of the Sun and other young stars.

ANCIENT ATMOSPHERES

We can now distinguish between two types of atmospheres — those that are primitive, representing material captured from the solar nebula with only minor fractionations, and those that are secondary — produced by the degassing of the material that accreted to form the planet. The first type of atmosphere is found in the outer solar system; the inner planets exhibit the second type. To give a quantitative illustration of the extent to which the light gases are deficient on the inner planets, one can imagine adding hydrogen and helium to the Earth until the abundance ratios of these elements to silicon were equal to the ratios found in the Sun. The resulting planet would have a mass approximately equal to the mass of Saturn.

Since we have good evidence that the composition of our planet's atmosphere has changed with time, it is natural to ask whether at some earlier stage it could have been strongly reducing, like the atmospheres we now see in the outer solar system. In other words, did the Earth and the other inner planets capture atmospheres from the solar nebula as they formed? And if they did, were the compositions of these atmospheres similar to those we now find in the outer solar system? With time, these planets would inevitably have moved from a highly reduced to an oxidized condition, since hydrogen will escape from their small gravitational fields and their warm surface temperatures ensure that water will be available to provide a source of oxygen. They must have begun with no free oxygen in their atmospheres, but were these atmospheres ever as reducing (or hydrogen-rich) as Jupiter's?

For many years this seemed the most likely scenario, but the evidence that was cited in its favor has become less compelling in the wake of new discoveries. An early argument involved the distribution of abundances of the noble gases in our present atmosphere. Compared with the cosmic abundance pattern, the noble gases are clearly deficient on Earth, and this deficiency seems to be mass-dependent, such that helium is the most depleted gas, then neon, then argon, etc. This pattern was taken as evidence that the Earth once had an atmosphere with its full complement of volatiles, but a "catastrophic event" — perhaps a strong solar wind during an early unstable and flaring stage of the Sun's history (T-Tauri phase) — swept away the gases, removing the lighter ones most efficiently. It seemed that the noble gases left us a record of this event, since they are chemically inert, and (except for helium) are too heavy to escape from the Earth's gravitational field.

But now we know that this same abundance pattern is found in meteorites and in the atmosphere of Mars, although the atmosphere of Venus is distinctly different. It thus seems that the fractionation processes were acting in the solar nebula *prior* to the formation of the planets, and affected all of the solid bodies in the inner solar system in different ways. We have also learned that the T-Tauri phase of solar history was probably not cataclysmic enough to have blown gases away from the surfaces of planets.

To reconstruct the Earth's early atmosphere, we must therefore turn this argument around. This fractionation of the noble gases in the solar nebula was presumably accompanied by a fractionation of other gases as well. Therefore, the *maximum* amount of hydrogen that the Earth could have captured can be calculated by using the neon in the present atmosphere as an index. In other words, if we assume that neon was captured, the cosmic hydrogen-to-neon ratio would give us the maximum value for early atmospheric hydrogen. This turns out to be about 10 millibars, or 1/100 of our present atmospheric density. Such an atmosphere would be lost by escape in less than 10,000 years.

Methane and ammonia would have had abundances 1,000 times smaller than that of hydrogen. Ammonia is particularly unlikely as a long-term atmospheric constituent, since this small amount would be destroyed in less than 40 years by solar ultraviolet light. Ammonia would also have been out of equilibrium with crustal rocks.

Thus, the only hope for a highly reducing early atmosphere would seem to reside in the possibility of producing it by degassing from the early Earth, either by internal melting processes caused by radioactivity or by external processes — bombardment by meteorites and comets that incidentally may have contributed reduced volatiles themselves. The first of these possibilities requires the presence of some reducing agent in the upper mantle. Free iron has been suggested as a candidate for this role, assuming that this early degassing took place prior to formation of the Earth's core. Some investigators are unhappy with this picture, however, arguing that the core should have occurred as the planet itself formed, since the energy of accretion would have been sufficient to initiate the melting of iron once the embryo-Earth attained 10% to 25% of its present size.

At this stage of our knowledge, there seems no way to rule out a transient, early atmosphere rich in hydrogen, methane, and carbon monoxide and containing ammonia. Because of the difficulties referred to above, however, interest is shifting toward the possibility that the early atmosphere of our planet was only weakly reducing, consisting of a mixture of nitrogen, carbon dioxide, carbon monoxide, water, and a few percent hydrogen. Such an atmosphere would provide a sufficient greenhouse effect to keep the Earth warm even if the Sun were 25% less luminous (a model which has some support) during that period than it is today. There is no need for ammonia or some other reduced gas to provide this effect as long as the partial pressure of CO_2 is on the order of 200 millibars. This large amount of atmospheric CO_2 would gradually diminish as a result of rock-weathering and the consequent production of carbonates.

Everything that we have described for the Earth should apply to Mars and Venus as well. Why then has our planet turned out so differently from our neighbors? Let us return to our basic criteria for planetary differences — size and distance from the Sun.

We first consider distance from the Sun. Suppose we could move the Earth to the position occupied by Venus. What would happen? The increased intensity of sunlight would cause the mean temperature of our planet to rise. Model calculations show that this increase in temperature would cause increased evaporation of sea water and would lead to a larger amount of water vapor in the atmosphere that could increase the greenhouse effect and the mean surface temperature further, leading to more evaporation, etc. In other words, the Earth's climate would go into a positive feedback loop that would lead to a condition known as a runaway greenhouse. The end result would be that the oceans would boil, putting all of the surface water into the atmosphere. The atmosphere itself would be so hot that there would no longer be a cold trap to confine water vapor to lower levels. Photodissociation by ultraviolet light would then become very efficient, and the hydrogen atoms resulting from this process would escape into space.

This is one explanation for the absence of water on Venus. The process we have just described would have occurred on that planet shortly after it formed. The oxygen left over from H_2O dissociation would have combined with the crust and available volatiles; we see one of the results in the dense carbon dioxide atmosphere that now blankets the planet (fig. IV-5). Plausible as this scenario seems, it has not yet been rigorously proven. A demonstration that deuterium is enriched on Venus would provide a strong supporting argument. If indeed there were once oceans that boiled with the subsequent

Figure IV-5.— *Venus, the planet closest to the Earth in size, has retained a significant atmosphere of carbon dioxide. Cloud features indicate a significant atmospheric dynamics.*

escape of hydrogen, the heavier mass of deuterium should have led to a net enhancement of this isotope in residual hydrogen compounds over geologic time. The alternative to this picture is that Venus had no (or very little) water from the beginning, a result again of its proximity to the Sun; the temperature at which Venus formed was so hot that water could not have condensed. This alternative seems less satisfactory in view of recent models that explain the bombardment histories of the inner planets by impacts from meteorites and comets that had formed in colder parts of the solar system. These impacts would necessarily inject additional volatiles such as water into the atmosphere of an evolving planet. While the visible results of this bombardment seem only superficial, it is likely that even during the early periods of planet formation, there was opportunity for ample mixing of materials from various parts of the solar nebula.

Thus, if the Earth were much closer to the Sun than its present position, it would be too hot for liquid water to be stable on its surface. And without water, life as we know it cannot survive, and probably cannot even originate.

What if Earth were farther from the Sun? This situation is more complex. A change in the composition of our planet's atmosphere could lead to an enhanced greenhouse effect, resulting in temperatures above the freezing point of water, even at the distance of Mars. Indeed, model calculations show that if Mars itself had an atmosphere in which the partial pressure of CO_2 was equal to the total pressure of our own atmosphere, the mean surface temperature on Mars would be above $0°$ C. This result has been used to explain the presence of sinuous channels and other landforms on Mars that seem to provide evidence for the action of running water on the planet's surface at some time in the distant past — perhaps 3.5 b.y. ago.

If Mars had a sufficiently dense CO_2 atmosphere during its early history, it would have been warm enough to allow liquid water to exist, and that water could have cut the channels we observe today. Thus, it seems possible that Mars may have had an early history much like that of the Earth. Yet we do not see evidence that there were ever lakes or oceans on Mars. It looks more as though the water came out in periodic floods, but never accumulated in large basins. Perhaps conditions were simply not that stable. Was the atmosphere never dense enough to do more than raise the temperature closer to the freezing point than we find it today?

In any case, the water present produced its own demise. Dissolving atmospheric CO_2 caused weathering of the rocks that led to the formation of carbonates, thereby reducing the amount of CO_2 in the atmosphere. This negative feedback would ultimately lower the surface pressure and the surface temperature. Abetted by other factors, such as the loss of nitrogen by escape, this process led to the point where liquid water could no longer exist.

But we must now consider our other basic planetary characteristic: size. What if Mars were a larger planet? We can return to our hypothetical experiment of moving the Earth to the position of Mars. In that case, more carbon dioxide could have been released initially, and the greater amount of tectonic activity associated with the larger heat engine in the bigger planet would provide a much more efficient means of recycling that gas. Perhaps a planet the size of Earth or slightly larger would be able to maintain a reasonably warm climate at the Mars distance from the Sun, provided that it could maintain a sufficiently large amount of carbon dioxide in its atmosphere. This in turn might heat up the tropics of a planet still largely frozen. Weathering would proceed very slowly, since much of the crust would be protected by ice. Whether life could originate and persist on such a planet is a matter of speculation. But these considerations do show that an Earth-like planet could exist at a greater range of distances from its central star than we would have concluded had we required that the planet have an atmospheric history identical with our own.

Still farther from the Sun, we must consider other liquids and, hence, other kinds of life. Ammonia is often suggested as an alternative to water as a medium for alien types of biology. We should expect to find such environments on the surfaces of satellites of the outer planets, since the planets themselves do not have solid surfaces that would allow ammonia (or any other liquid) to collect.

The best example of an object that might meet these criteria is Titan, the largest satellite of Saturn. This is the only satellite in our solar system known to have a substantial atmosphere; this atmosphere contains both nitrogen and methane. The problem with Titan is that it is too cold to be very interesting. The surface temperature has been shown to be below $-175°$ C by both spacecraft and Earth-based measurements. If there is a liquid on Titan's surface, it is liquid methane or liquid nitrogen, not ammonia; ammonia would be solid.

The development of alien life at these low temperatures seems unlikely in view of the slowness of chemical reactions under these conditions. Yet some very interesting chemistry is taking place on Titan, for we can detect traces of reaction products in the atmosphere. In addition to CH_4, C_2H_2, C_2H_4, and C_2H_6 that had been detected by Earth-based telescopes, the Voyager spacecraft succeeded in identifying N_2, H_2, HCN, C_2N_2, HC_3N, C_4H_2, C_3H_4, and C_2H_8. Furthermore, the atmosphere is charged with a brownish photochemical aerosol that may include polymers of one or more of these substances. Since hydrogen can escape from Titan, fragments of hydrocarbons that are produced by UV irradiation or charged particle bombardment in the satellite's atmosphere are free to combine to form more complex substances. Here, we have a highly evolved atmosphere that has remained reducing, since oxygen is safely trapped as water ice in Titan's interior.

This lack of liquid water makes the current chemistry on Titan fundamentally different from the chemistry on the primitive Earth. But the chemistry occurring in the Earth's atmosphere in its early history may well have been very similar to what we find on Titan today, making further investigation of this object particularly appealing. It depends, of course, on how reducing our early atmosphere was. At this stage of our ignorance, a mixture of CH_4, N_2, and H_2 can't be excluded.

The photochemical reactions are taking place in the satellite's upper atmosphere, which is some 80° C warmer than the surface. The reaction products will gradually settle out and be preserved in this cold trap (or dissolve in liquid methane or liquid nitrogen!). The early history of Titan may have been even more interesting, however. The nitrogen we now find in the atmosphere is presumably the result of the photodissociation of ammonia. In order for the ammonia to get into the atmosphere, Titan must have been much warmer. This would have led to much more methane in the atmosphere as well, so we can imagine an early atmosphere even denser than the present one, which has a surface pressure of 1.6 bars. How warm was the satellite during this early period? How long did this time last? What kinds of chemistry occurred? Here in this cold, alien environment, we find ourselves confronting the same questions we have considered on Mars.

We can find in this example another important property of water that makes it well-suited as a liquid medium for life. When ammonia is photodissociated, the reaction products are nitrogen and hydrogen, neither of which protects the ammonia from further photodissociation. It thus seems problematical whether one can ever have an environment in which liquid ammonia is stable. In contrast, the oxygen produced by the breakup of water can act to shield the water while also providing a potential source of chemical energy (far more accessible than nitrogen) for evolving life. An alternative would be to provide a UV-protective smog layer, such as Titan in fact seems to possess. The trick is then to maintain a warm enough surface to have liquid ammonia but a cold trap high in the atmosphere that prevents the ammonia from diffusing up to altitudes where the smog cannot protect it. On Titan, no such ammonia sea is present, but perhaps it is present somewhere far away.

The outer planets themselves are less promising. There is ample evidence for chemical reactions, particularly on Jupiter where we see a variety of colors among the clouds (fig. IV-6). The expected condensed ammonia, ammonium hydrosulfide, ammonium hydroxide, and water all preclude white clouds. Hence, the existence of colors indicates that more complex, nonequilibrium compounds are being formed. With solar ultraviolet light, lightning storms, bombardment by charged particles, and escaping internal heat all available as energy sources, we have a giant natural laboratory in which experiments bearing on the first stages of chemical evolution are continuously being performed. At the present time, these colored substances have not been identified. It is becoming increasingly evident that we shall have to probe Jupiter to solve this problem.

THE ANCIENT SURFACES

The recent exploration of the planets has given us a model of the early Earth, but what terrestrial evidence do we have of what our Earth looked like in the past? The most ancient metamorphosed sedimentary rocks now known are those at Isua in western Greenland. They were deposited roughly 3.8 b.y. ago as sediments carried by water into a volcanic basin to which volcanos contributed solids and

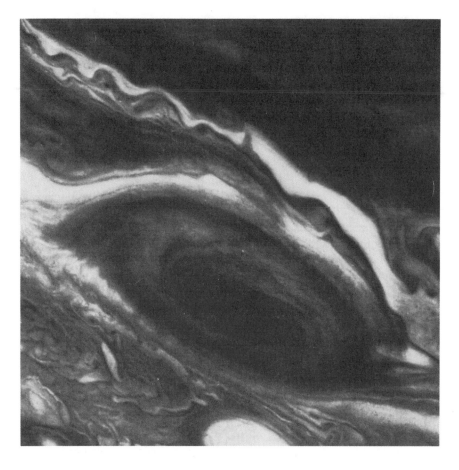

Figure IV-6.– *The giant red spot of Jupiter and the complex cloud patterns suggest a natural laboratory for chemical evolutionary processes.*

probably hydrothermal solutions as well. Some metamorphosed sedimentary units at Isua approach the composition of normal present-day sedimentary rocks, which bear conclusive evidence of an atmosphere that contained sizable quantities of carbon dioxide. Little can be said about the oxygen content of the atmosphere at the time of deposition of the Isua sediments. There is a fair amount of evidence that the oxygen content of the atmosphere more than 2 b.y. ago was substantially less than today. Therefore, there must have been less oxygen at the time of deposition of the sediments at Isua. The atmosphere at that time was certainly not highly reducing; in fact, the bulk composition of the Isua rocks is surprisingly similar to that of younger rocks in equivalent settings. In this sense they give eloquent testimony to the notion that 3.8 b.y. ago the Earth had already settled down to a regime that is quite similar to that of the present day.

Elemental carbon is present in these sediments. The Isua sediments have been heated to such high temperatures since deposition that virtually no extractable organic compounds remain. At present, a biological origin for carbon in the Isua rocks seems unproven. Life could have started before the time when Isua was formed, perhaps much closer to the birth of our planet. A search for sedimentary rocks older than 3.8 b.y. is obviously needed.

Several continents are known to contain 3.5-b.y.-old sedimentary rocks, and these ancient areas may contain enclaves of even older rock units. Areas in Australia and southern Africa, and an area in central Greenland that is currently covered by ice may turn out to be particularly promising targets in the search for more ancient rocks.

We believe that our Earth is about 4.6 b.y. old. At present we are forced to look to other bodies in the solar system for hints as to what the early history of the Earth was like. Studies of our Moon, Mercury, Mars, and the large satellites of Jupiter and Saturn have provided ample evidence that all of these objects were bombarded by bodies with a wide variety of sizes shortly after they had formed. This same bombardment must have affected the Earth as well. The lunar record indicates that the rate of impacts decreased to its present low level about 4 b.y. ago. On the Earth, subsequent erosion and crustal motions have obliterated the craters that must have formed during this epoch. Since it is generally believed that life on Earth began during this period, the bombardment must have been part of the environment within which this event occurred.

Perhaps the most significant aspect for our consideration is the realization that some of the impacting objects were large enough to punch through the crust of Earth and formed large basins that could be flooded by lava. These would be the terrestrial analogues of the large circular maria on the Moon, the basins on Mars, and the ringed structures on Jupiter's moons — Callisto and Ganymede. This is a reminder that the early history of the Earth was more turbulent than a simple volcanism model would suggest.

We have seen how interesting and important it is to find rocks from the earliest possible times in the Earth's history. Yet, it has been much easier to find ancient rocks on the Moon, where many samples have been dated at 4 b.y. or more. Unfortunately, these ancient lunar rocks do not tell us about the Earth's primitive atmospheric conditions, since the Moon evidently possessed no long-enduring, substantial atmosphere, even in those early times. The lunar rocks are grossly deficient in volatile elements and compounds compared with the Earth, suggesting that they were thoroughly degassed but that the gases escaped rapidly into space.

There are large regions on the surface of Mars that exhibit crater densities similar to those found on the lunar highlands, the oldest region of the lunar surface. Although there is still some dispute about absolute chronologies, this similarity in the distribution of impact-craters has led several investigators to suggest that these regions of the Martian surface are probably as old as the comparable areas on the Moon. Thus, it seems reasonable to expect that on Mars as on the Moon, rocks with ages greater than 4 b.y. should be reasonably abundant, provided one were to go to the right region of the planet to look for them.

The great difference between Mars and the Moon is that Mars has an atmosphere. This atmosphere has apparently developed from an inventory of volatiles very similar in composition to the one that formed the atmosphere of Earth. Indeed, there is every reason to expect that the first steps in the development of the atmospheres on these two bodies were essentially identical. Although we would expect hydrogen to escape more rapidly from the low gravitational field of Mars, this may have been partially compensated for by the lower temperature of the exosphere on this more distant planet. Thus, we can suggest that if Earth ever had a strongly reducing atmosphere, Mars probably did too.

This probability lends a special piquancy to the search for ancient rocks on Mars. If we could find such rocks in a suitably protected setting, we would have an opportunity to test the possibility that the early Martian atmosphere was strongly reducing by examining the mineral assemblages that the rocks contain. Our conclusions would then be very useful in helping us to evaluate the composition of the atmosphere on the primitive Earth.

But we can go further than this. The surface of Mars is also marked by many different examples of fluvial erosion. Once again the timing of the events that led to these landforms is controversial. But it is probably conservative to say that the youngest of the numerous floods took place at least 3.5 b.y. ago. In other words, liquid water was evidently available on some kind of intermittent basis for the first billion years of Martian history. The controversy that still exists centers on the issue of how much later than this there might have been epochs when water flowed on Mars. For our immediate purpose, it doesn't matter; 1 b.y. is long enough!

As already stated, our current view of the development of inner-planet atmospheres suggests that Mars, Earth, and Venus all began with a very similar volatile inventory. Models for the early Martian climate indicate that a dense CO_2 atmosphere could have melted ice by a greenhouse effect. The morphological evidence that liquid water once flowed on Mars seems to substantiate both of these points. The special significance of this picture of primitive Martian history is apparent as soon as we ask what was happening during the first billion years of our own planet's history. There is almost no direct evidence available to answer this question. But the recent discovery of stromatolites dated by three different methods to be 3.5 b.y. old indicated that life had originated, evolved, and become firmly established on our planet within the first billion years. If this happened on Earth, why not on Mars?

There seems no way to exclude this possibility. We might be more comfortable if there were a record of ancient seas and lakes on Mars — proof that the presence of water was more than a series of very transient events. But the evidence for such smooth landforms is much more likely to disappear under the shifting sands of the wind-blown Martian terrain than is the high relief associated with individual stream beds. In fact, there are many examples of craters with diameters greater than 30 km that once contained standing water. These "lakes" existed at the same time as the large fluid channels. Periodic wetting and desiccation and/or freezing could help to concentrate prebiotic material, as has often been stressed in considerations of the origin of life on Earth. We are thus confronted with the arresting possibility that since life originated on Earth within the first billion years of its history, and since conditions on Mars were probably similar to those on Earth during this period, there is an excellent chance that life originated on that planet too. We can easily see why and how such life could have died out in the ensuing millenia, leading to the negative results obtained by the Viking investigations (fig. IV-7). But even if there is no life anywhere on Mars today, there seems to be good reasons for returning to Mars to look for evidence of early life forms.

This will require a careful search in "the right places," as has been required on Earth. A Viking-style lander mission has virtually no chance of success in such an endeavor; a manned mission or a sophisticated system of rovers with capabilities for sample return will be required. It will be an expensive and difficult task, but the rewards would be so great that a considerable effort is justified. To find one more example of the origin of life, to know that this

Figure IV-7.— *NASA's commitment to the exploration of the planets is most dramatically seen in Viking Martian lander.*

remarkable property of matter is indeed something that springs forth from natural causes wherever the conditions are right — these goals are surely worth the effort and expense.

A few remarks of caution may be of value. Now that we have a basis for comparison of the early history of the Earth's atmosphere with the present state of neighbors Mars and Venus, the conventional wisdom seems far from secure. Instead of the strongly reducing atmosphere rich in hydrogen — the methane, water, and ammonia mix of two decades ago — the best guesses now suggest a mixture still clearly reducing, but much less strongly, under the dominance of water with the oxides of carbon. No doubt the inferences are sensible, but are they unique? We need to recall that the comparison among planets must also include the history of the Sun. The solar models almost certainly require that the Sun slowly brighten; in the time since Earth formation the solar input has risen by a third or so, the same effect as a change in Earth orbit by 15%. So the old Earth and the present Earth are themselves two planets at different distances, so to speak. It is true that the solar inputs to Venus, Earth, and Mars are in the ratio of about 4:2:1; this difference is much larger than that from solar evolution. But is the input change to be neglected? Did it make some difference to the origin of life, a difference less important once life is vigorous, perhaps because of some feedback effect of life itself? The topic can serve to remind us that the early history of the planets is still only inferential, hardly part of a secure understanding. We should take care that the present best view does not jell into dogma.

SUGGESTIONS FOR FURTHER READING

Billingham, J., Ed.; Life in the Universe, NASA CP-2156, 1981.
Goldsmith, D.; and Owen, T.: The Search for Life in the Universe, Benjamin/Cummings Co., Inc., Menlo Park, California, 1980.
Walker, J. C. G.: Evolution of the Atmosphere, Macmillan, New York, 1977.

Plate.V.Part.III.

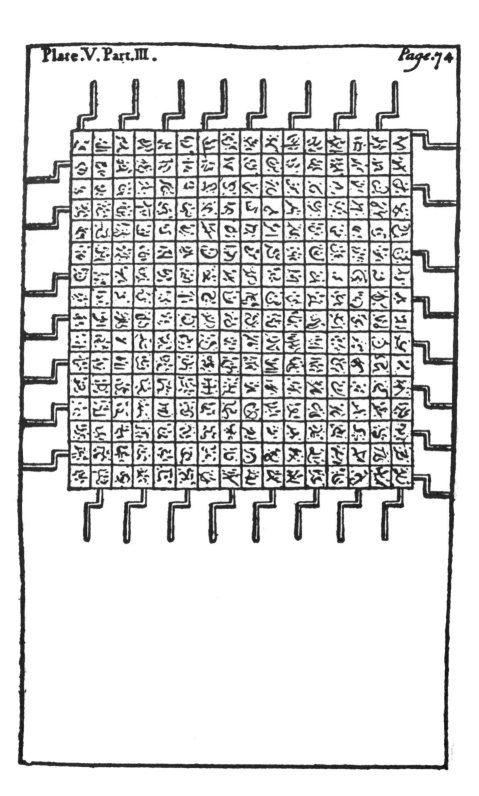

V. THE LABORATORY EXPERIENCE

INTRODUCTION

In chapter III we traced the history of the Earth back through time by means of the rock record, yet it stops at 3.8 b.y., and we know the Earth to be 4.5 b.y. old. We sought clues to the early stages of the Earth's formation in chapter IV in our reconstruction of the evolution of the solar system, especially of the Earth, from the interstellar gas and dust, yet we still found great gaps. One theme which runs like a thread throughout our previous chapters, however, is the role of organic molecules in the universe.

A brief glimpse into the nature of organic matter in our biosphere thus seems in order. It is necessary to spend some effort at a rather more technical level than in most of this book to discuss what are chiefly chemical questions.

In the 19th century, as the microscope was perfected, the cell was discovered. All organisms including ourselves are composed of cells. The main function of a cell is to grow and divide. By the use of dyes, various parts of the cell were recognized: nucleus, mitochondria, chloroplasts, and centrioles. It was a great age of observation;

A drawing made by an unknown artist to illustrate an early edition of Gulliver's Travels. Dr. Lemuel Gulliver [in "Gulliver's Travels"] reported on this device, which he saw in the National Academy of Laputa.

"It was Twenty Foot square ... The Superficies was composed of several Bits of Wood ... all linked together by slender Wires ..." On them "were written all the Words of their Language in their several ... Declensions, but without any order .... The Pupils at his Command took each of them hold of an Iron Handle ... and giving them a sudden Turn, the whole Disposition was entirely changed ... the Professor shewed me several Volumes ... already collected, of broken Sentences, which he intended to piece together ... to give the World a compleat Body of all Arts and Sciences ..."

The same logic applies to the origin of life, although we are looking for a less random hierarchy of processes, and a less elaborated system of selection.

Provided by Phil Morrison, Cambridge, Mass. 1983

the biology of microscopic life received a powerful stimulus from the rise of medical bacteriology.

It was left to the 20th century to analyze the cell into its chief chemical components: proteins, nucleic acids, fats, and carbohydrates. Of these chemical species, proteins were soon discovered to be giant molecules made up of thousands of atoms. If proteins are gently broken down, they fall apart into amino acids; there are only 20 different amino acids in all life. Thus, a protein can be described as a word string in an alphabet of 20 letters (the amino acids). It was found as well that the nucleic acids (DNA and RNA) were strings over an alphabet of four letters (nucleotides) (fig. V-1). In the case of DNA, the letters (or molecules) are adenylic acid (A), guanylic acid (G), cytidylic acid (C), and thymidylic acid (T). In the case of RNA it is A, G, C, as above, but uridylic acid (U) instead of thymidylic acid (T). The number of combinatorial possibilities are more than astronomical. Assuming we have formed a string of amino acids 100 letters long, how many different ones could be present? There are 20 different possibilities for the first member of the string and 20 for the second and so on. Therefore, there are $(20)^{100}$ different proteins, a number large past imagining.

The DNA in *E. coli* is found in a single molecule which is about 1 mm long. There are about 3 million nucleotide base pairs in one molecule of DNA. The different protein strings realized are encoded in the sequence of bases in the double-stranded molecule of DNA.

A protein molecule called RNA polymerase transcribes the sequences within the DNA molecule into RNA molecules, called messenger RNA molecules. An RNA molecule then enters a protein-synthesizing machine, which is best compared to a molecular tape recorder, in which the RNA molecule is read and the output is a sequence of amino acids.

Since there are four bases (A, G, C, U) in the RNA alphabet that can be used to code for amino acids, it can be seen that three bases ($4^3 = 64$) are required to accomplish this code: neither one base ($4^1 = 4$) nor two bases ($4^2 = 16$) would provide a unique code for the 20 amino acids. The representation of the sequence of amino acids of proteins by a sequence of nucleotides in RNA is called the genetic code (table V-1). In table V-1 the first base in the triplet is listed in the first column, the second base is listed along the row, and the third base is listed in the last column. The 64 triplets are thus

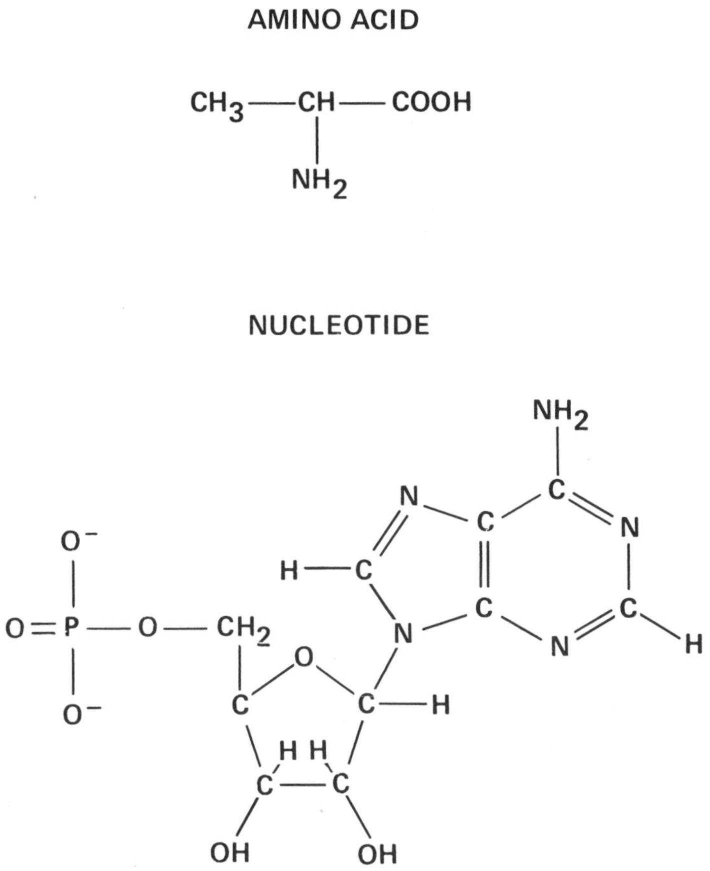

AMINO ACID

$$CH_3 \text{---} CH \text{---} COOH$$
$$|$$
$$NH_2$$

NUCLEOTIDE

Figure V-1.— *Shown are representative examples of an amino acid, a nucleotide. Amino acids are the building blocks of the proteins, while the nucleotides are the building blocks of the nucleic acids. A complete set of these molecules used in biology can be found in the appendix.*

related to the 20 amino acids. For example, the triplet GGG stands for the amino acid glycine. In summary, the "central dogma," which can be stated DNA → RNA → Protein, dictates that the synthesis of proteins is controlled by the nucleic acid and that the genetic code describes the relationship of the simple molecules that make up the strand of nucleic acids and proteins to each other. But where did these simple molecules come from and how were they organized?

TABLE V-1.– GENETIC CODE

Bases					
First	Second				Third
	G	C	A	U	
G	gly	ala	glu	val	G
	gly	ala	glu	val	A
	gly	ala	asp	val	C
	gly	ala	asp	val	U
C	arg	pro	gln	leu	G
	arg	pro	gln	leu	A
	arg	pro	his	leu	C
	arg	pro	his	leu	U
A	arg	thr	lys	met	G
	arg	thr	lys	ile	A
	ser	thr	asn	ile	C
	ser	thr	asn	ile	U
U	trp	ser	term	leu	G
	term	ser	term	leu	A
	cys	ser	tyr	phe	C
	cys	ser	tyr	phe	U

EXTRATERRESTRIAL EVIDENCE

The study of meteorites — stimulated by the fall of one big unusual carbonaceous meteorite on Murchison, Australia, in 1969 — has allowed close examination of extraterrestrial material. Numerous laboratories have shown that the Murchison meteorite contains many organic compounds that appear to have been synthesized extraterrestrially by nonbiological processes. One of the first classes of compounds studied was the amino acids. As compared to the 20 amino acids found in proteins, a much larger number (over 100) have been estimated to be present in carbonaceous chondrites.

(The amino acids in the Murchison meteorite are optically inactive, unlike those in proteins, showing that both handed forms of a given amino acid were present in roughly equal proportions; see fig. V-2.) In addition to these fundamental building blocks of life, other important classes of organic molecules have been identified. These include heterocyclic bases, hydrocarbons, fatty acids, hydroxyacids, etc. These solvent-soluble materials represent at most 30% of the carbon found in these meteorites (or about 0.5% of the total meteorite weight). The remaining carbon is predominantly present as a solvent-insoluble phase.

At this point it is important to remark that while the types of organic compounds found in meteorites are consistent with those expected to have served as precursors for the biochemicals of terrestrial life, we do not know the details of their synthesis on the meteorites. We cannot be sure that either the environments in which

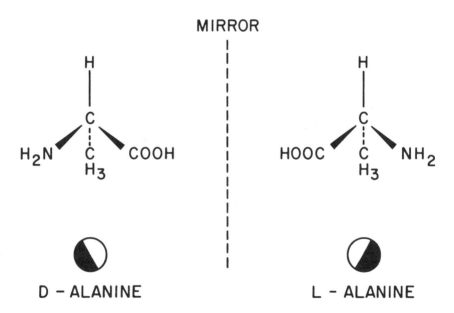

Figure V-2.– *An important property of carbon is one that allows the formation of two forms with identical composition – one form is the mirror image of the other, just as the left hand is the mirror image of the right hand. Thus there are two forms of alanine, L (levo-) the left handed and D (dextro-) the right handed versions. The L-amino acids predominate in terrestrial biology.*

the meteorites originated or the organic matter synthesized were the same as those on the primitive Earth. However, the complex suite of organic molecules found in carbonaceous meteorites provides a new basis for our understanding the phenomenon of chemical evolution.

As mentioned in chapter IV, the widespread occurrence of organic compounds — not produced by any life form — in the cosmos and within our solar system confirms the expectation, based on elemental abundances in the stars, that organic chemical evolution is one natural consequence of the evolution of matter in the universe. But organic chemical evolution is inextricably intertwined with the evolution of environments, be they interstellar clouds, meteorite parent bodies or planets; its progress toward the origin of life may be terminated at different stages depending on the physical and chemical constraints imposed by the environment.

As noted in previous chapters, the Earth's atmosphere, oceans, and crust during the first 500 million years are difficult to define in any detail. Too little is known to fix the actual time of accretion, the heterogeneity of the accreting materials, the state of the Sun, and the quantity of short-lived radioisotopes that could have influenced the thermal structure of the early Earth.

In many respects our knowledge of the early Earth is much like our knowledge of the early solar system. It is model-dependent and relies on the reconstruction of an environment by extrapolation from a record preserved in (but deciphered only in fragmentary fashion from) lunar rocks, meteorites, and remotely discernible features of Venus and Mars and of very ancient rocks and sediments of the Earth. As more of the record is unveiled, new evidence is revealed, new interpretations arise, and models undergo revision. Thus, all that can be done at present is to define a rather wide range of possible compositions for the early atmosphere and oceans and to suggest the implications of atmospheres and oceans within this range for the problem of the origin of life. These models have provided the basis for experimental simulation studies.

THE EXPERIMENTAL ERA

Laboratory efforts have been extensive and have provided much of the insight we have gained into the origins of life on Earth.

Because of the uncertainty of the nature of the primitive atmosphere, experimentation has explored a range of plausible models from those whose gas composition is strongly reducing to those with a more oxidizing composition.

Strongly Reducing Atmosphere

We will now explore that model which takes as its premise that the initial atmosphere of the early Earth was rich in hydrogen compounds, the one which has received most attention in the last generation.

According to the Oparin-Miller-Urey paradigm, a highly reducing atmosphere consisting of methane, ammonia, and water − all hydrogenous − prevailed on the primitive Earth. Passage of energy in various forms through this hypothetical atmosphere produced the reservoir of organic molecules from which life evolved.

The first successful prebiotic amino acid syntheses were carried out using his reducing gas mixture of CH_4, NH_3, H_2O (or CH_4, NH_3, H_2O, H_2) and an electric discharge as an energy source. The result was a large yield of amino acids (the yield of glycine alone was 2.1% based on the amount of carbon present), together with hydroxy acids, short aliphatic acids and urea. One of the surprising results of this experiment was that the products were not a random mixture of organic compounds, but rather a relatively small number of compounds were produced in substantial yield. In addition the compounds produced were, with a few exceptions, of biological importance.

The special mechanism of synthesis of the amino and hydroxy acids was further investigated. It was shown that the amino acids were not formed directly in the electric discharge but were the result of solution reactions of smaller molecules produced in the discharge, in particular, hydrogen cyanide and aldehydes. These reactions were studied subsequently in detail and the equilibrium and rate constants of these reactions were measured. These results show that amino and hydroxy acids could have been synthesized at high dilutions of HCN and aldehydes in a primitive ocean.

Ultraviolet light acting on this mixture of gases is not effective in producing amino acids except at very short wavelengths

($<$1500 Å) and even then the yields are very low. The low yields are probably a result of the low yields of HCN produced by UV light. If the gas mixture is modified by adding gases such as H_2S or formaldehyde, then reasonable yields of amino acids can be obtained at relatively long wavelengths ($<$2500 Å) where considerable energies from the Sun are available. The H_2S absorbs at these longer wavelenths and is photodissociated to H and HS. The H atoms have a high velocity ("hot atoms") and react with the CH_4 and NH_3. It is possible, but not demonstrated, that HCN and other molecules are produced, which then form amino acids in the aqueous part of the system. In experiments simulating thermal environments, pyrolysis of CH_4 and NH_3 gives low yields of amino acids.

A second model of a reducing atmosphere with less hydrogen would consist of CH_4, N_2, and traces of NH_3 and H_2O. This atmosphere is more consistent with current models of the primitive Earth, though still consisting of hydrogenous compounds. Large amounts of NH_3 would not have accumulated in the atmosphere because of photodestruction and also because the NH_3 would dissolve in the ocean.

This mixture of gases is quite effective with an electric discharge in producing amino acids. The yields are somewhat lower than with CH_4, NH_3, and H_2O but the products are more diverse. Hydroxy acids, short aliphatic acids, and dicarboxylic acids are produced along with the amino acids. Ten of the 20 amino acids that occur in proteins are produced directly in this experiment. Methionine is obtained if H_2S is added to the mixture, while cysteine, another sulfur containing amino acid, was found in the photolysis of CH_4, NH_3, H_2O, and H_2S. Phenylalanine, tyrosine, and tryptophan can also be synthesized under putative prebiotic conditions. Thus, only the basic amino acids − lysine, arginine, and histidine − have not been produced in prebiotic synthesis of amino acids. There is no fundamental reason that the basic amino acids cannot be synthesized, and this problem may be solved before too long.

Mildly Reducing Atmospheres

The geochemists, especially W. Rubey, were not happy with the proposal by Urey that the early atmosphere was composed of methane, ammonia, and hydrogen. They favored a model for the atmosphere provided from volcanic outgassing. It was dominated by carbon dioxide, nitrogen, and water vapor; Abelson argued that "volatiles from outgassing interacted with the alkaline crust to form an ocean having a pH 8–9 and to produce an atmosphere consisting of CO, CO_2, N_2, and H_2." A series of experiments were initiated in a mildly reducing atmosphere.

CO, N_2, H_2 — Electric discharges acting on this mixture of gases are not particularly effective in amino acid synthesis, but HCN is produced in significant amounts. Glycine is produced in fair yield, but only small amounts of higher amino acids are formed. However, formaldehyde, which is important in the prebiotic synthesis of sugars, is obtained in large amounts.

CO_2, N_2, H_2 — The CO_2 is more oxidized than the CO, but the excess H makes it a reduced mixture. As with $CO + N_2 + H_2$, the amino acid synthesis is quite low with electric discharges unless H_2/CO_2 ratio is $\gtrsim 2$. In this case glycine is produced in fair yield, but again very few of the higher amino acids are formed.

CO, H_2 — This mixture is used commercially in the Fischer-Tropsch reaction to make hydrocarbons in high yields. The reaction requires a catalyst, usually Fe or Ni supported on silica, a temperature of 200°–400°C, and a short contact time. Depending on the conditions, aliphatic hydrocarbons, aromatic hydrocarbons, alcohols, and acids can be produced. If NH_3 is added to the $CO + H_2$, then amino acids, purines, and pyrimidines can be formed. The intermediates in these reactions are not known, but it is likely that HCN is involved together with others.

CO, H_2O — Electric discharges are not effective with this mixture, but UV light that is absorbed by the water (≤ 1849 Å) results in the production of formaldehyde and other aldehydes, alcohols, and acids. The yields are fair. The mechanism seems to involve splitting the H_2O to H + OH with the OH converting the CO to CO_2 and the H reducing another molecule of CO.

The amount of hydrogen needed in the synthesis outlined above, except in the CO, H_2O experiment, is still very high and it does not fare well with present ideas about the atmosphere of the early Earth.

Nonreducing Atmosphere

If the early Earth's atmosphere were dominated by the gases, carbon dioxide, nitrogen, and water, then the Miller-Urey experiment would not be relevant to the origin of terrestrial life. The reduction of carbon dioxide and nitrogen would have to have taken place by means other than molecular hydrogen. Recall that reduction means the acceptance of electrons by a molecule or ion. Thus, an organic molecule in water, upon accepting an electron, gains a net negative charge which is neutralized by a proton donated by the water molecule itself.

If the early atmosphere were not hydrogen-rich, the reduction of carbon dioxide could only be carried out if another supply of electron donors were available. A conceivable source of electron donors on the early Earth would be iron – the ferrous ions.

In 1960, Getoff irradiated an aqueous solution of ferrous sulfate and carbon dioxide with light ($\sim$2600 Å) and got a yield of formaldehyde of approximately 1%. These observations should be considered in the light of recent proposals that the reducing conditions on a primitive Earth were to be sought in the abundance of ferrous iron in the crust rather than in the amount of hydrogen in the atmosphere. The problem of photochemically reducing N_2 and NH_3 is currently under active investigation. Low yields of NH_3 form from N_2 in the presence of the metals Mo, Fe, and Ti when irradiated by long-wavelength UV light. The reduction of carbon dioxide and nitrogen may have been possible on a primitive Earth if the right electron donors were present. This opens up new experimental vistas in the studies of the origin of life, especially in the synthesis of the amino acids. While the above discussion has dealt predominantly with the synthesis of amino acids, we would be remiss if we neglected the laboratory synthesis of the other organic species we have encountered in this text.

As pointed out many times in the text, the nucleic acids play a fundamental role in terrestrial biology. In addition to phosphorous in the form of phosphate, the nucleic acids require both sugars and certain nitrogenous molecules, the purines and pyrimidines. There is ample laboratory evidence to show that these molecules are produced from the reactions of the simple molecule HCN in water. Of the purines, adenine (see Appendix) is the major one produced from

HCN oligomers. Of the pyrimidines, cytosine is formed directly by the reaction of cyanoacetylene and cyanate, while uracil is the final product using either HCN or cyanoacetylene as the starting material. These are only examples from an extensive scientific literature.

Sugars and Nucleosides

The self-condensation of formaldehyde was a likely route to sugars on the primitive Earth, the formaldehyde being formed by the action of electric discharges or UV light on a mildly reducing atmosphere. This condensation is inhibited by HCN which reacts rapidly with formaldehyde. Therefore, the synthesis of sugars probably was delayed until the bulk of the HCN had condensed or was hydrolyzed. It is not clear how the relatively few sugars which have the central role in contemporary living systems (ribose and glucose) were selected from the very complex mixture of compounds which is formed from formaldehyde.

Lipids

The prebiotic formation of lipids has not been extensively investigated, but the limited experiments which have been performed suggest that lipid-like materials might have also formed spontaneously on the primitive Earth.

Biosynthetic Pathways as a Guide to Prebiotic Chemistry

Since the primitive metabolic pathways probably evolved from prebiotic syntheses, some steps in the current metabolic pathways may be "chemical fossils" of reactions which occurred on the primitive Earth. Consequently, a proposed prebiotic pathway gains validity if it can be correlated with a contemporary biosynthetic process.

There is a good correlation between the contemporary biosyntheses of purines and pyrimidines and some of the steps in their proposed prebiotic syntheses.

As an example, the decarboxylation of the nucleotide of orotic acid is one of the steps in the biosynthesis of the nucleotides of uracil and cytosine. This same decarboxylation is a key reaction in the prebiotic synthesis of uracil from HCN. In addition, aspartic acid, the starting material for the biosynthesis of orotic acid, is produced in a variety of prebiotic experiments. It would not have been a major change for early life to evolve a system for the biosynthesis of orotic acid from the readily available aspartic acid once the low supply of preformed orotic acid limited the growth of primitive life forms.

INORGANIC ASPECTS

While much of the work in the study of chemical evolution and the origins of life have dealt with the formation, polymerization, and interaction of important organic molecules, there is an awareness on the part of most students of the field that inorganic chemistry was undoubtedly of fundamental importance in the processes responsible for the origins of life on Earth. Since it is known that all life on Earth now requires the metal ions for its chemical function, a number of scientists have questioned at what stage such fundamental processes became important. Early in the history of this field of study, Granick suggested that the first organization of preprotoplasm could be a primitive energy-conversion unit that could perform the elementary processes of photosynthesis and respiration; that this unit originated within the domain of some common minerals; that the minerals that contain metal ions served both as coordinating templates and catalysts for various reactions; and that around this unit were formed organic molecules that gradually became organized into units of ever-increasing complexity. Thus, biosynthetic chains developed in a stepwise fashion. The metal catalysts of minerals became modified into the metalloenzymes; in these new complexes the same metals would become more efficient.

In a similar vein, Bernal had earlier suggested in his book "The Physical Basis of Life" that clays were sites upon which organic molecules could be concentrated and react with each other.

In recent years, studies of the role of metal ions and minerals in prebiological chemistry have shown promising results. As noted earlier, the reduction of CO_2 to formaldehyde and methane has been accomplished by the interaction of UV light with an aqueous solution of CO_2 and ferrous ions. In preparation for the Viking mission to Mars, the reduction of CO_2 was also observed to occur on silicate surfaces. The reduction of nitrogen to ammonia by titanium dioxide has been accomplished, again by using UV light.

In laboratory experiments, biomonomers have been synthesized in the presence of clays. Investigations have shown that clays affect the formation of amino acids and nitrogen heterocycles from CO and NH_3 at temperatures of about 300°C. In addition, biomonomers can be adsorbed onto clays. This adsorption provides an excellent mechanism of concentration to facilitate subsequent chemical reactions. The clay- and/or metal ion-mediated oligomerization of biomonomers has also been demonstrated. Through this mechanism polypeptides and oligonucleotides have been formed in higher yield or with longer strands than in the absence of these inorganic components. Thus, inorganic chemistry may have played an important role in the emergence of life on Earth.

POLYPEPTIDES, POLYNUCLEOTIDES, AND THE BEGINNINGS OF NATURAL SELECTION

The transition from a mixture of organic molecules to an organized system that is capable of reproducing itself, represents the most puzzling problem in the study of the origin of life. We know that contemporary cells rely on proteins, very complex molecules, to catalyze specifically almost all biological reactions, including the replication and translation of nucleic acids. The proteins are themselves the products of the translation process, and their synthesis is in turn dependent on the presence of preformed nucleic acids. The origin of the genetic process thus appears to be a chicken-and-egg problem; which came first — proteins or the coded nucleic acids?

Not everyone agrees that studies of the origin of self-replicating systems should concentrate on nucleic acids and proteins. Some researchers suggest a variety of simple, alternative self-replicating systems. Cairns-Smith, for example, proposes an entirely inorganic genetic system based on cation substitutions in clays. The major idea is that clays could not only adsorb and catalyze reactions between organic molecules but that they could, like DNA, replicate. If we now suppose that, as in the case of DNA, the possibility exists of an error of replication or mutation, the replicating clays would evolve! At present, however, the self-replicating systems we know best involve molecules that resemble proteins and nucleic acids. The remainder of this section, therefore, is concerned with studies of these molecules.

The Contemporary System

A genetic apparatus is an essential requirement for all living things on Earth. It is by means of the genetic material that living organisms are able to store, express, and upon reproduction, transmit to their progeny the information for all of the capabilities which they possess. Cellular life forms usually store genetic information in double-strand DNA polymers (fig. V-3), though some viruses make use of RNA instead. As mentioned earlier, the information is coded in the sequence of the nucleotides in such a way that each of the 64 possible trinucleotides codes for one of the 20 acids to be incorporated into a protein (table V-1), or codes for a stop signal to terminate protein synthesis. The two strands of the nucleic acid are held together by relatively weak (hydrogen) bonds, made specific by a unique and essential feature of the conformation of the four nucleotides — adenylic acid hydrogen-bonds specifically with thymidylic acid, while guanylic acid pairs only with that of cytidylic acid (fig. V-4). These interactions are referred to as the Watson-Crick pairing rules. Partly because of these unique pairing specificities, either strand can serve as a template for the synthesis of the other, given free energy, in the enzyme-catalyzed process known as replication. In this fashion, one double strand can yield two new double-stranded molecules, one for each of the progeny after cell division.

The expression of genetic information requires that the information in the DNA be converted into protein. This is accomplished in two steps. First, in a process called transcription, one strand is copied by an enzyme, yielding a complementary strand of RNA (messenger RNA). This new strand then serves as the template for the synthesis of protein. The process of protein synthesis is termed translation because the nucleic acid "language" is now translated into

SIDE VIEW

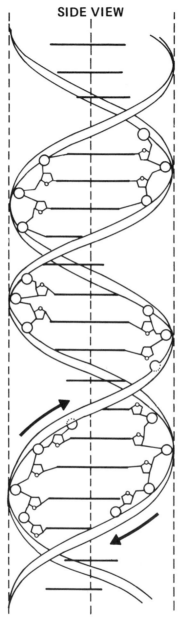

Figure V-3.— *DNA (deoxyribonucleic acid) consists of two polymers linked together by pairs of purine and pyrimidine bases. Of the four types of bases, adenine can pair with thymine and guanine can pair with cytosine (Watson-Crick pairing rules: A-T; G-C).*

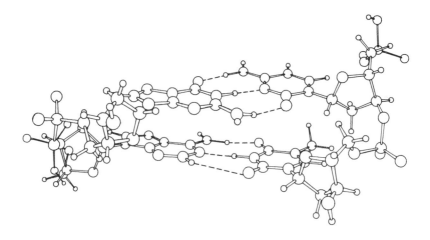

Figure V-4.– *The two strands of nucleic acids are held together by hydrogen bonds which are specific for the four nucleotides. A "double helix" is thus created and these interactions are referred to as the Watson-Crick pairing rules.*

protein "language." The "dictionary" which establishes the rules by which this translation is accomplished is called the genetic code. The genetic code, shown in table V-1, was thought to be entirely universal until quite recent experiments showed that in mitochondria, one of the stop signals actually codes for one of the amino acids, tryptophan; another amino acid, methionine, has two codons rather than only one. While these exceptions are rather minor variations on a major theme, they do emphasize the fact that isolated systems can evolve slightly different codes.

Which Came First: Proteins or Nucleic Acids?

Several people have proposed that the first genetic system was composed of proteins alone. They point out that it is much easier to synthesize and polymerize the amino acids than the nucleotides. Furthermore, the resulting polymers have some catalytic ability; in the contemporary cell, proteins play the major role as catalysts. Nucleic acids, on the other hand, are not usually thought of as having catalytic activity, although they do act as templates for replication and transcription (and a "bringing together" of reactants is certainly one important function of many catalysts). In what follows, we will examine first the systems that contain only proteins, then those that contain only nucleic acids, and finally the combined systems.

Proteins first— The nonordered polymerization of amino acids to form analogs of proteins (called peptides if the chains are short) has been demonstrated in the laboratory under a wide variety of experimental conditions. The energy needed for this reaction can be provided in a number of ways, one of which is heat. Polyamino acids (sometimes called proteinoids) have been produced by heating mixtures of amino acids to about 180°C. These temperatures are greater than the boiling point of water; however, in a few experiments similar products were found at temperatures below the boiling point of water, after very long periods. Clays also catalyze the reac- when alternately wetted and dried. The formation of peptides on clay is further catalyzed by a simple peptide, i.e., histidyl-histidine. This suggests that the first enzyme-like molecules may have been very simple peptides.

Another approach is to provide the energy needed for synthesis of peptide bonds through the use of other energy-rich chemical species called condensing agents. Still another approach is to use amino acids which have been "activated" prior to reaction. These experiments are convincing evidence that amino acids readily enter into combinations with one another, but these experiments do not address the problem of reproducible organization into specific sequences which would participate in self-replication.

The main difficulty with the "protein first" hypothesis is that it does not seem possible to formulate a plausible scheme for protein self-replication based on known properties of amino acids and peptides. One type of proposal postulates a complementary pairing of amino acids on two chains, analogous to base-pairing in nucleic acids. Complementariness might depend on size, charge, hydrogen bonding, or some combination of these properties. Such schemes are certainly possible in principle, but there is no experimental evidence for them.

Another group of proposals suggests that a family of peptides forms a cycle in which each member of the cycle catalyzes the synthesis of one or more other members of the cycle. It has been shown that the synthesis of certain peptide antibiotics of well-defined sequence is brought about by a group of specific proteins (enzymes) without the help of nucleic acids. However, the peptide antibiotics that are synthesized are relatively simple and molecules of this size would not be capable of catalyzing the synthesis of complex specific proteins. Certainly, the enzymes involved are very complex and are themselves synthesized with the help of nucleic acids. We do not think, therefore, that this system is a good model for protein self-replication. It is dangerous to be dogmatic about general schemes of self-replicating peptide cycles, but we suspect that they all suffer from the same problem: simple peptides lack sufficient specificity, while large peptides are too hard to make so that it would be impossible to close the cycle. Furthermore, even if one self-replicating cycle existed, it is hard to see how it could evolve to greater complexity.

Nucleic acids first— There are a number of theories in which polynucleotide replication is proposed to have preceded the synthesis of ordered polypeptides. Proponents of such theories emphasize that nonenzymatic complementary replication of polynucleotides seems plausible in light of the known interactions between nucleotide bases. We shall see that there is already a substantial body of experiments supporting the idea that a preformed polynucleotide can direct the synthesis of a complementary oligonucleotide according to the previously described Watson-Crick pairing rules. The required "preformed" nucleic acid may itself have arisen originally in a non-ordered joining reaction of mononucleotides; such reactions are already known. The polymerization of nucleotides, just as in the related reaction of amino acids to form a polypeptide, effectively requires the removal of one molecule of water for each addition of a nucleotide to the growing chain. Therefore, it is not surprising to find that successful polymerizations have usually employed drying conditions, the addition of "condensing" reagents, or the removal of water in a prior step to form an activated monomeric nucleotide. The addition of a catalyst, such as a metal ion, can increase the yield of polynucleotides quite appreciably. It has been shown that divalent metal ions enhance the formation of oligoadenylates from an *activated* derivative of adenylic acid. In the case of lead ion a 56% yield of oligomer was formed, while in the absence of this metal ion the yield of oligomer was about 4%. It has recently been shown that simple catalysts can be important in more sophisticated sequence-copying reactions, and certainly contemporary replication enzymes have all been found to contain a metal ion at the active site.

Template-directed polynucleotide synthesis—The central reaction responsible for the stability of inherited characteristics is nucleic acid replication, while a major source of genetic variation is the inaccuracy of this process. This had led many researchers to postulate that nucleic acid replication, in which a preformed polynucleotide template directs the synthesis of a new complementary strand, was the first "genetic" process of the primitive Earth. This theory is appealing, but it should not be accepted as dogma.

Under certain conditions, organized if unusual helical structures can form between polynucleotides and the complementary monomeric nucleotides or nucleosides. It has been shown that if certain activated condensing agents are added to energize the system, the monomers can join to form short oligonucleotides. These reactions have established that the Watson-Crick pairing rules apply to these systems too in a nonbiological setting. A detailed analysis of the products of early experiments revealed a startling structural difference between this chemical condensation and the normal enzymatic reaction. (The predominant internucleotide linkages in the chemical product are $2'-5'$ rather than $3'-5'$ (fig. V-5), but in contemporary cells, it is the later linkage which is found almost exclusively.)

Several years ago, it was shown that the Zn^{2+} and Pb^{2+} ions are effective catalysts for energized template-directed synthesis yielding oligomers of guanylic acid with chain lengths in excess of 30. (The Pb^{2+} ion gives predominantly $2'-5'$-linked products, while the Zn^{2+} ion gives mainly $3'-5'$-linked products.) The Pb^{2+} reaction has an error rate of about 0.1 in the presence of a "wrong" base, while the Zn^{2+} reaction has a much lower error rate of 0.005. Thus, the Zn^{2+} catalyzed reaction produces products with the "usual" linking and with an accuracy comparable to the accuracy required by a non-enzymatic replicating system.

A more recent modification with a slightly different activating group produces a comparable preference for the "usual" $3'-5'$ linkage and comparable accuracy, without requiring Zn^{2+}. True

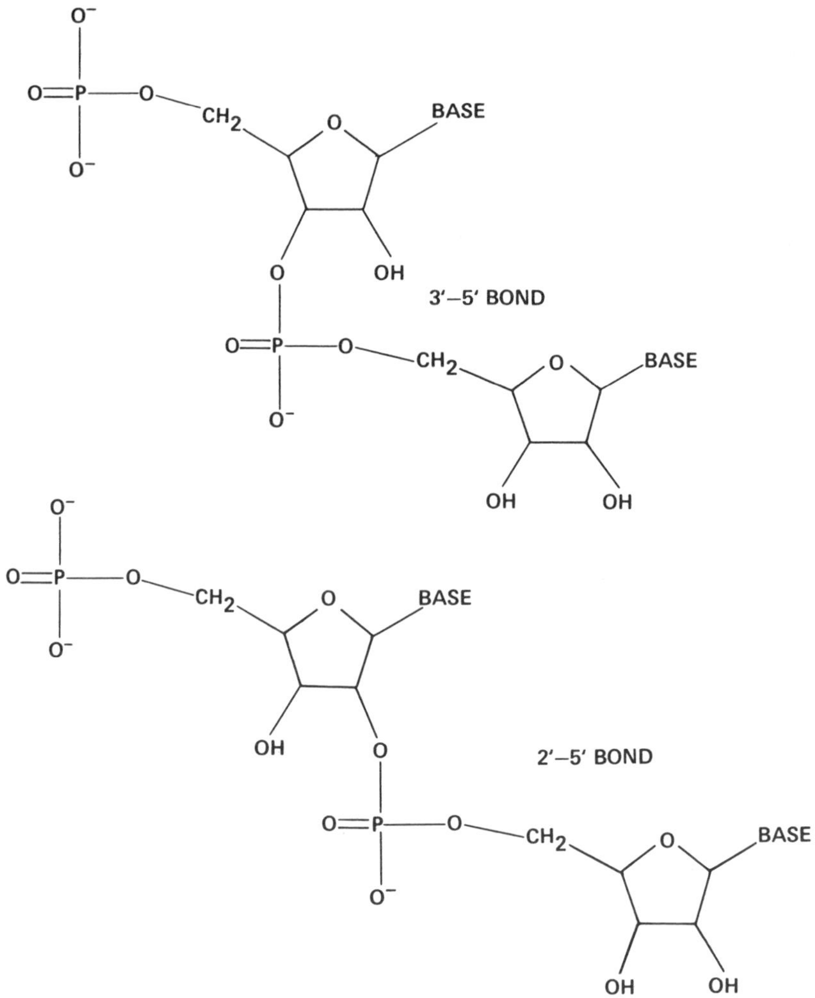

Figure V-5.— *There are two possible ways to form phosphodiester bonds in the process of building a polyribonucleotide from the monomers. Biology utilizes only one of these linkages in RNA. That linkage is the 3'-5' one illustrated above.*

nonenzymatic replication of a nucleic acid with a mixed base sequence through several generations of "offspring" has not yet been demonstrated experimentally.

Thus, the question of whether nucleic acid sequences could evolve by natural selection cannot yet be answered in a prebiotic system, but experiments using biological enzymes were done to address this question. Replicating systems consisting of the enzyme QB-RNA polymerase and certain small RNA substrates have been used to demonstrate molecular evolution in the test tube. One set of experiments started with an RNA substrate that absorbs a dye, ethidium bromide, and is then unable to replicate efficiently. By allowing the RNA to replicate repeatedly in the presence of gradually increasing concentrations of the dye, a new RNA substrate was generated which was no longer inhibited, because it no longer bound the dye-stuff so tightly. This system does not show *self*-replication, because each round of synthesis required the addition of new enzymes. However, experiments like these do show that RNA molecules can adapt by natural selection to bind or reject specific organic molecules, and this is relevant to discussions of the origins of the genetic code.

Recent experiments demonstrate for the first time that nucleic acids themselves might cause interesting reactions to occur. An RNA strand was observed to "snip out" a portion of its own sequence, apparently without help from enzymes. This cannot really be called catalysis, since the molecule acts on itself, and does so only once for each molecule, but it does suggest that nucleic acids may be capable of at least a few specific catalytic-type reactions. However, in the absence of additional evidence that polynucleotides are able to function as catalysts, one cannot feel confident that nucleic acids could do enough interesting chemistry to "go it alone."

As with amino acids (fig. V-2), nucleotides possess optically active configuration. Observations that D-nucleosides react more efficiently than L-nucleosides on a nucleic acid template made of D-nucleotides, suggests that nucleotide chains that are made up of monomer with the same enantiomeric configuration can undergo template-directed replication, while "mixed" oligomers cannot. Of course, either all D- or all L-oligomers would replicate equally well, so this result shows only that the components of the primitive nucleic acids must either all have been L-, or all have been D-isomers.

It does not explain why only D- instead of L-nucleic acids are important in biology.

Proteins and nucleic acids together— Since neither proteins alone nor nucleic acids alone seem likely to be able to account for all of the genetic properties needed for self-replication, the alternative is to consider the development of a combined system of both proteins and nucleic acids. This requires a coupling between the two kinds of molecules in the form of at least a primitive genetic code.

While speculations as to how genetic coding might operate even predated elucidation of the structure of DNA in 1953 by Watson and Crick, an understanding of the essential nature of the genetic material greatly stimulated the desire to understand how it is expressed. As information accumulated in the 1950s and 1960s about the molecular mechanisms of transcription, translation, and the coding process, another question began to emerge. Why, for example, is UUU a code for phenylalanine? There has perhaps been more speculation about the basis for the origin of the genetic code than any other aspect of molecular biology, and to enumerate and discuss them all is impossible in a short review. The theories fall into two groups: (a) Were the genetic code assignments based on some relationship (perhaps affinities) between amino acids and nucleotides, or (b) were they the result of random processes? The idea that the code is based on chance evolutionary processes, implies that we are not likely ever to understand the basis for the origin, so that experimentalists have necessarily been concerned with the first theory. Evidence for a physicochemical basis for the code, however, has not been abundant. Experimentalists have shown that mononucleotides have differential affinities for polybasic amino acids, but these affinities relate more to the self-associative properties of the mononucleotides than to code-related specificities. However, recent work has shown a preferential affinity of certain polyamino acids, polylysine for A-T rich DNA, and polyarginine for G-C rich DNA. In a similar fashion, it has been shown that the aromatic amino acids (tryptophan, phenylalanine, and histidine) have different affinities for polyadenylic acid. This information was important in showing that, at the monomer level, selectivities do exist.

Investigations have demonstrated that there are variations in affinities of nucleotides for amino acids affixed to a column material, but these variations were not clearly related to the code. Similarly, studies have shown a differential uptake of nucleotides and amino acids into detergent droplets called micelles, but code-related correlations were not evident from this work either. In 1976 the first clear-cut, code-related correlations between hydrophilic (water loving) and hydrophobic (water hating) properties of amino acids and nucleotides were reported. In addition, other researchers showed that a number of additional properties, including hydrophobicity, polarity, and bulkiness were also correlated between amino acids and nucleotides. Thus, very weak interactions can be detected between amino acids and nucleotides, but whether these interactions provide enough selectivity for a translation process is not yet clear.

The demonstration of a weak, specific interaction between an amino acid and a nucleotide does not put us much farther forward unless we can couple it to an efficient peptide synthesis reaction. Many mechanisms have been proposed for simple systems that might be capable of translating nucleic acid sequences into peptides of protein sequences, but no experiment has ever shown actual translation in the absence of the complex components of the contemporary cell. A few experiments have shown a slight influence of nucleic acid on the yield of condensation of a single amino acid, with variation of yield as either the amino acid or nucleic acid is changed, but these experiments are not convincing evidence of a translation effect. As a result further experimental verification is in order.

There is no paucity of suggestions for the origin of protein synthesis and of the genetic code, but there are few suggestions that are both chemically explicit and believable. Some papers are rather philosophical and do not lead immediately to a testable experiment. Others are explicit but the chemistry appears to be inconsistent with present knowledge. Actually the requirements of an efficient peptide synthesis system can be stated quite simply. The reacting amino acids, perhaps attached to specific oligonucleotide carriers, need to be lined up on the template, such that adjacent amino acids are held *close enough together and in the proper orientation* to allow reaction. The amino acids need to be in a sufficiently reactive form, so that the peptide bond formation is a spontaneous reaction, but not so reactive that hydrolysis competes with peptide bond formation. In addition, the mechanism has to be recursive, i.e., repetition of the process should result in a gradual elongation of the peptide chain.

Proposed translation models— Although most of the models that have been suggested for the first translation apparatus closely resemble the contemporary process of reading a linear message with triplet adapters, a number of other models have to be kept in mind as possibilities. Examples include systems involving adapters reading a single base rather than triplets, mechanisms with direct bonding of amino acids to polynucleotide templates, and various feedback linked systems. These tend to suffer from the problem of explaining how the contemporary genetic system could have evolved from such a different mechanism. On the other hand, some of these alternative models tend to be more easily evaluated experimentally. For instance, recent experiments have demonstrated that amino acids attached to single nucleotides give an enhanced yield of peptide formation when the nucleotides are lined up on a complementary strand of polymer. There clearly remain many other novel mechanisms to be discovered and suggested.

It must be kept in mind that the problem may not yet be solvable because some vital information may not have been discovered. Additional knowledge of the three-dimensional structure of ribosomes and transfer RNA is beginning to shed more light on how the complex translation apparatus operates in the contemporary cell. Structural studies of nucleic acids can still yield some surprises. It seems likely that increasingly detailed knowledge will suggest new possibilities for prebiotic studies of the translation mechanisms.

In spite of the caveats that we must acknowledge, the correlations of properties and affinity data between amino acids and nucleotides, while certainly leaving us far from final answers, at least suggest that discernible patterns exist in the coding mechanism and give hope that primitive translation can be elucidated when sufficient data are available.

MEMBRANES

We can now ask what kinds of structures could coevolve with a replicating system that might enhance the ability of the system to incorporate the functions described above, and to evolve toward the structure we accept as a living cell. A key component of all cells is the cell membrane, which allows cells to maintain an internal milieu different from the external environment in the composition and concentration of compounds. The main structural component of all membranes − with rare minor variations − is the lipid bilayer. It is as fundamental a structure in cells as the DNA double helix, in that it constitutes the basic permeability barrier which delimits the cells and controls their interaction with the environment.

In prebiotic experiments, some alternatives to lipid bilayers as permeability barriers have been proposed, including coacervates, microspheres of proteinoid, or other heterogeneous polymers, micelles, and surfaces of minerals. All of these show some ability to selectively concentrate or retain some of the molecules characteristic of living cells. However, none of them is as efficient as lipid bilayers, and in addition it is difficult to account for how these could evolve into the lipid bilayer membranes of modern cells.

In the absence of an understanding of exactly how membranes or other structures became coupled to genetic mechanisms, it is best to remain undogmatic about which structures were most important, or about whether genetic mechanisms developed before or after the structural components.

It is useful at this point to list some properties of lipid bilayers that may be relevant to the development of replicating systems on the prebiotic Earth. These properties fall into subcategories that include physical, chemical, and supramolecular aspects of lipid organization in aqueous environments. Only one class of lipids, the phospholipids (fig. V-6), is generally involved in the formation of membranes. It is a striking property of most phospholipids that in an aqueous environment they form stable bilayer structures that typically close to form vesicular membranes. Since phospholipids have been synthesized under plausible prebiotic conditions and have been demonstrated to form vesicles, it is assumed that lipid bilayer vesicles were present on the prebiotic Earth.

The next question concerns how such structures might contribute to prebiotic evolution. The chemical properties of lipid bilayers include a highly charged surface and a nonpolar interior of the bilayer. These two properties represent an almost totally unexplored area for research in prebiotic evolution, and suggest a number of new

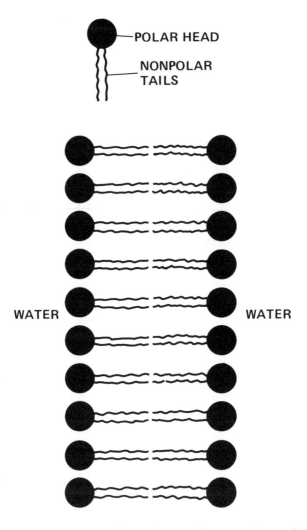

Figure V-6.– *The membranes of most living cells are made up of (a) phospholipids. They spontaneously form a lipid bilayer (b) in water as illustrated above.*

research directions. For instance, it is likely that the charged surfaces, like those of clays, have catalytic properties, and this possibility should certainly be investigated.

The hydrophobic moiety also holds considerable interest. All contemporary light energy transducing systems depend on a nonpolar membrane phase to embed specialized pigments and enzymes, and to provide a barrier for the electrochemical gradients produced by the pigment-enzyme systems. It is reasonable to assume that pigment molecules, formed under prebiotic conditions, would partition into the nonpolar phase of lipid bilayer membranes and offer primitive light energy-trapping functions.

A significant physical property of the lipid bilayer is its relative permeability to various ionic and molecular species. In contemporary cells, the bilayer is understood to be the major barrier to free diffusion of water-soluble substances. The special permeability properties of cell membranes must be attributed to their other major constituent, the proteinaceous channels and enzymes that permit or drive ion and chemical transport. Recently, small peptides (e.g., valinomycin) have been shown to form specific ionic channels in a lipid bilayer.

Energy Transduction by Early Membranes

We can now go on to discuss possible contributions of membrane structure to evolution of energy transduction systems on the prebiotic Earth. The energy necessary to generate and maintain the organization of early life forms has usually been assumed to have come from chemical reactions of compounds synthesized by prebiotic processes. There is, however, another source of energy, which has largely been overlooked. Given the existence of lipid bilayer-enclosed vesicles, concentration gradients across the lipid bilayer would arise if the environment should change after the vesicles have formed. Chemical reactions favored by the milieu inside the vesicles would also set up concentration gradients. Concentration gradients of charged solutes could generate diffusion potentials and the flux of one charged species across the lipid bilayer could be coupled to the flux of any other charged species. If amino acids are polymerized inside a vesicle, this would give rise to an amino acid gradient, as would the breakdown of the polymer. In a similar fashion, many chemical reactions would change the acid/base nature of the vesicle interior. These could couple chemical reactions in the vesicle interior to transport processes. For example, many organic weak acids and bases permeate lipid bilayers easily in their undissociated form and would accumulate or be depleted inside vesicles if pH gradients existed.

Gradients of other ions such as sodium, potassium, and calcium can be modulated by carriers or channels. Relatively simple organic compounds are known which act as carriers for these ions, and it is possible that analogous compounds could have arisen abiogenically. It should be noted, however, that in contemporary cell membranes channels rather than carriers appear to be the main mediators of specific ion permeabilities.

Closed membrane compartments play, of course, a crucial role in the two most important contemporary energy conversion systems — photosynthesis and respiration. Their evolution has been extensively discussed and will not be further pursued here. However, cells presumably existed before either of these mechanisms evolved and our main purpose is to discuss the problems of evolution from abiogenic organic matter to protocells. The speculations advanced here are necessarily based on our knowledge of present-day living cells. Unfortunately, fossils tell us little about the chemistry of Precambrian life.

Thus, we have been taken full circle back to the Precambrian. We have traced our knowledge from the fossils dated 3.5 b.y. through our speculations about the early Earth from 4.5 b.y. until 3.5 b.y. We must note here that this gap of almost 1 billion years is slowly being filled by our experiments in the laboratory and by our expanding knowledge of our Earth and solar system. We have therefore deemed it necessary to outline in the concluding chapter the next steps in our quest to close the gap and elucidate the nature of the origins of life on Earth.

SUGGESTIONS FOR FURTHER READING

A. G. Cairns-Smith: Genetic Takeover. Cambridge University Press, London, 1982.

W. Day: Genesis on Planet Earth. Talos Press, East Lansing, Mich., 1979.

R. E. Dickerson: Scientific American, vol. 239, no. 3, 1978, pp. 70–87.

J. G. Lawless, C. E. Folsome, and K. A. Kvenvolden: Organic Matter in Meteorites. Scientific American, vol. 226, no. 6, 1972, pp. 38–46.

S. Miller and L. Orgel: The Origins of Life on Earth. Prentice Hall, Englewood Cliffs, N.J., 1974.

VI. RECOMMENDATIONS

INTRODUCTION

As a field of active scientific inquiry, the study of the origins of life is clearly in its infancy. By its very nature, it is multidisciplinary, requiring contributions from astronomers, biologists, chemists, geologists, physicists, and many others.

We have seen that the gaps in our knowledge of the steps from the nonliving to the living are numerous. Among these gaps are (a) a solar system formation with its accumulation of raw materials, (b) the synthesis of the life-forming monomers, such as the amino acids, nucleotides, and lipids, (c) the condensation of these monomers into useful polymers such as proteins and nucleic acids, (d) the sequestering of these materials into droplets of proteinoid or membrane-like structures, and (e) the development of a chemical memory (the genetic code) to pass on to the progeny the information acquired.

Throughout the text we have shown the partial answers to the many questions we have asked about organic chemical evolution and the origin of life, yet much remains to be learned. Thus, it behooves us to identify those unanswered questions. The following discussions, therefore, represent an attempt to identify these issues with the provisions that (a) the compilation is not intended to be complete, but rather, it represents important areas of future research identified by a specific group of people, and (b) progress in any one of these areas might change the whole direction of research in the field.

Professor Harold Urey was a remarkable American scientist who as much as any other man opened up the field of origins of life to modern study.

SOLAR SYSTEM

The record of the early history of the solar system may be preserved in the organic and inorganic matter of the comets. Present evidence is consistent with two possible scenarios for the origin of the organic molecules sublimating from the cometary nucleus. Either they represent the "frost" of interstellar molecules that condensed onto those interstellar grains that have later accreted into comets, or they represent the "snows" that condensed onto silicate grains during the cooling phase of the presolar nebula. If they represent the frost of interstellar molecules, we must obtain a more detailed understanding of what this frost is. To accomplish this we propose to elucidate the major processes (stellar, circumstellar, interstellar, etc.) involved in the formation of gaseous organic molecules detected in interstellar space. For example, is the formation of much more complex solid polymeric organic molecules possible in interstellar dust-ice grain surfaces? How could they be detected?

In addition, we need to acquire new and improved receivers at millimeter and submillimeter wavelengths, and new, accurate, and large telescopes to collect these wavelengths. Unavailable funds in this area would likely set the United States back in its worldwide lead in the study of the interstellar molecules during the 1980s. Therefore, a greater effort should be exerted to convince the Government of the significance of these research tools.

However, if the comets represent the snows that condensed during the primitive solar nebula, then the organic and inorganic constituents would record for us the composition of the early solar system.

We can't go on much longer without knowing more about comets. Even if these objects did not bring biomonomers to the surface of the primitive Earth, they may have been a major source of the volatile elements from which these compounds were later formed. We need to study comet nuclei with a rendezvous mission that brings a spacecraft close enough to permit detailed measurements of the nucleus and inner coma. The program should build toward a capability for sample return and interception of a large, active comet that is approaching the inner solar system for the first time.

Another possible record from which to study the composition of the early solar system is meteorites. Consideration of models for the origin of solid bodies in the solar system, and data obtained from the study of meteorites raise a number of important issues.

Differentiation and hydrothermal processes occurred on small bodies very early in the chemical evolution of the solar system. Similar processes took place on the primitive Earth, but the chronology and the consequences of these processes on the thermal, geochemical, and atmospheric evolution of the Earth's prebiotic environment are major unknowns, and require elucidation. Insofar as processes operating on meteorite parent bodies can be generalized or related to similar processes on the Earth, then continued study of their manifestations in meteorites will contribute significantly to understanding the environment in which life originated.

The hypothesis that an interstellar cloud of dust and gas provided the spawning ground for the entire solar system is generally accepted. Models based on this hypothesis hold promise of providing a coherent framework for understanding the origin of the solar system and of, ultimately, life itself. In this context, relationships should exist between the organic matter in interstellar clouds, comets, and carbonaceous meteorites; thus, research efforts aimed at elucidating the nature of the relationship should be strongly encouraged. Contributions to the confirmation or denial of models for the origin of bodies in the solar system should become a major objective of organic chemical evolution research.

Planetesimals resembling meteorites and their parent bodies are believed to have supplied the building blocks for formation of the planets. Carbonaceous meteoritic material would have contributed significantly to the crust, mantle, and inventory of volatiles of the primitive Earth. How these contributions influenced its physical-chemical evolution and, therefore, the setting in which life originated should be of major concern in the study of the origin of life.

We still don't know where the carbonaceous meteorites came from; however, asteroids represent prime candidates. We do know that there are large numbers of dark asteroids whose reflectance spectra resemble those of carbonaceous chondrites. Is there a connection? Do these objects also contain organic compounds?

From the above considerations it is obvious that detailed studies of meteorites should be continued. And, a mission to the asteroid belt which could sample and return materials to Earth is important in our attempt to elucidate the history of the early solar system and, hence, to the origins of life.

Evidence bearing on the problem of the chemical setting for prebiotic synthesis may yet be forthcoming from comparative planetology. The information derived from the knowledge of our sister planets will provide constraints for the origin and development of our Earth.

The colors on Jupiter have long been thought to indicate the presence of organic chemical synthesis. Since the highly reducing atmosphere of Jupiter is consistent with some models of the Earth's early atmosphere, Jupiter may well represent a model for primitive Earth's chemistry. It is becoming increasingly evident that we are going to have to go there to find out. The Galileo Project will make a first step in this direction, but it is clear that more sophisticated explorations are required.

In the same vein, Titan, despite its low temperature, remains important because of its methane/nitrogen atmosphere, its red color, and the variety of evidence suggesting the presence of a photochemical smog. Because chemistry is taking place in this primitive reducing atmosphere today, and the products of these reactions are believed to be accumulating on the satellite's surface, a well-preserved record of carbon cosmochemistry undoubtedly exists. Thus, preliminary explorations of this environment are in order.

Within the context of a totally different planetary environment, there is still much to be learned from Mars after Viking. For example, judging from the ages deduced for some parts of the surface from densities of impact craters, there should be very old (older than 4 billion years) rocks on Mars. These rocks should presumably tell us whether or not Mars ever had a strongly reducing atmosphere.

In addition, detailed surface analysis of Mars may yet provide information about organic matter in protected environments, and information on the ages of various strata would be fundamental in understanding the epochs of liquid water. Finally, the mineralogy would help to answer two important questions. First, are the "missing volatiles" tied up in the form of carbonates, nitrites, nitrates, and sulfates? And second, laboratory studies using iron-rich clays appear to satisfactorily explain the results of the Viking experiments. Are these clays present, and if present what is their nature? Thus, we are presented with a comparative approach in assessing the early evolution of the solar system and the origin of life. We are adjusting our sights to the understanding of the planetary stage on which life enters as a player. Was the ancient Earth the only stage on which life could play? This is the deepest question that future research must answer. From comets to planets, NASA is the prime agency gathering the data necessary for providing an answer to this question.

THE EARTH

Studies of the origin of life require an accurate reconstruction of conditions and events on the early Earth. Investigations of those matters, once rare, are becoming more common as new techniques become available and as other developments in Earth science allow problems to be more clearly defined and profitably attacked. As this work proceeds, it must constantly be borne in mind that the early Earth was, quite literally, "a different planet." The contrasts between surface conditions on the early Earth and those on the modern Earth are nearly as large as the contrasts between the surface conditions presently found on the Earth and on Mars. It is not just a play on words, therefore, to speak of a "mission to the early Earth" just as we might speak of a "mission to Mars." While the latter involves the use of spacecraft and a journey of millions of miles, the former presents equal challenges (and promises equal scientific returns) in a journey across billions of years.

The space sciences thus have a special contribution to — and should have a special interest in — the studies of the earliest phases of Earth's history. The space scientist's avoidance of "geocentric bias," the unwarranted attribution of Earth-like characteristics to other planets, is perfectly appropriate to studies of the early Earth. Equally, the development of an accurate view of the earliest stages of Earth's history can provide crucial information regarding the origin and formation of planets generally, and can significantly constrain theories regarding the origin and development of the solar system.

Contrasts between the ancient and the modern Earth are nowhere greater than in biology, which has embodied a progression from no life, to primitive forms, to bacteria resembling those encountered today, to — after billions of years — single-celled plants and animals with biochemical systems like those found in higher life forms. This progression of microbes has caused events of planetological significance — the development of an oxygenic atmosphere and the deposition of vast mineral deposits being but two examples. Studies of the origin and early development of life, thus, simply cannot be separated from general investigations of the early history of the Earth as a planet. The geochemical record of the history of the volatile elements on Earth is the record of the history of life on Earth, and an understanding of that record is crucial to an accurate reconstruction of events in the early solar system.

The record from the early Earth has no parallel at later stages of Earth's history; close analogs may eventually be found on other planets, but no similar environments occur on the modern Earth. In a situation so without precedent, it will be necessary to constantly avoid unjustified extensions of present geochemical models. As it is acknowledged that the early Earth was without multicellular organisms or land plants, it must be recalled that this places *all* the primary productivity in the hands of microorganisms and creates a global ecosystem very different from any which has been considered for the past 0.5 b.y. As microbial ecosystems are then recognized as especially important subjects for study by planetary biologists, it must be recalled that the global impact of modern microbial ecosystems is buffered, perhaps powerfully, by the great mass of the biosphere which lies outside them.

The establishment of a research program at the interface between planetology, geology, and microbial ecology is, if anything, overdue. As it is welcomed and carried forward, the origins of its importance and the uniqueness of the problems it addresses must be kept constantly in mind. The best contributions will result from the arduous confrontation of all the evidence: biological, chemical, geochemical, geological, and astrophysical.

Such a research program must answer a number of questions. For example, can there have been any survival of prebiotic organic matter? How (chemically) might this prebiotic organic material be recognized? When did life first appear? To answer this question, the analysis of rock systems (e.g., Isua, Greenland; Swaziland, S. Africa; and Pilbara, N.W. Australia) is of prime importance. Studies should include those on organic geochemistry and isotope analyses, mineralogical and elemental analyses, and micromorphological and analogue community analyses.

Fossils are the records of historical events important in the origin and evolution of life. They exist in three forms: (1) embedded in rocks, (2) inherent in the complex, metabolic patterns of living organisms, and (3) recorded in the sequences of amino acids in proteins and in the sequences of nucleotides in RNA and DNA.

The knowledge we have gained from the rock record, in recent years, has improved measurably. In particular, photosynthesis, as represented by the photosynthetic bacteria, appears very early in the rock record. These findings are consistent with the hypothesis that autotrophic (or photosynthetic) organisms are the original life forms. This would upset the current conjecture that the original organisms were heterotrophic and hence, would change our perspective on the problem of the origin of life. The search for older rocks should be actively pursued.

To supplement the microfossil records, isotope fractionations of lighter from heavier isotopes of carbon and sulfur have been used. In the case of carbon the fractionation of isotopes has been interpreted as due to the fixation of carbon dioxide in photosynthesis by the Calvin cycle. The sulfur isotope fractionation is related to the reduction of sulfate to sulfide by various sulfur bacteria.

In the realm of metabolism, the fixation of carbon dioxide by various organisms by metabolic pathways other than the Calvin cycle and the related fractionations of isotopes has not been extensively studied. Examples are the fixations by Chlorobium (photosynthetic green sulfur bacterium) of carbon dioxide by a reverse citric-acid cycle, and the fixation by the methanogens of carbon dioxide by metabolic pathways still unknown. Thus, the interpretation of the isotope fractionation in the geological record is woefully weak on the biological side. A much more systematic study of carbon dioxide fixation is in order.

In the case of sulfur isotopes, the role of sulfide to sulfur metabolism in photosynthetic bacteria has been ignored in recent years in spite of the postulated early appearance of H_2S photosynthesis in the biosphere. Again this is evidence of a poor liaison between biologists and isotope geochemists.

Thus, the lack of a coherent attack on the problem is evident in carbon and sulfur isotopic fossil geochemistry and a close collaboration between the biochemist and the isotope geochemist is imperative to solve this important problem.

THE CELL

Studies in recent years have shown that the halophiles (e.g., Halobacterium halobium) create a proton gradient across their membranes after the absorption of light by bacteriorhodopsin. This is considered by many to be the most primitive photosynthetic model which we have available for study. As noted in chapter III the halophiles belong to the archeabacteria. These organisms which include the methanogens have very different cell membranes and coenzymes, and a protein-synthesizing machinery which has properties intermediate between procaryotic and eucaryotic cells. A current conjecture is that the archeabacteria branched off from other bacteria about 3.4 b.y. ago. Therefore, a detailed comparison of the two should push our knowledge of the biological record back. A coherent and systematic use of sequences of amino acids and nucleotides to clarify this split has just begun. The continual investigation of these sequences will be important in elucidating the nature of early life for us.

Among the kinds of questions or problems that should be addressed is a general one — the description and reconstruction of the universal common ancestor. At what stage in evolution did the entity exist? Was it preprocaryotic? Why was there a *common* ancestor? Was it chance? Was it necessary? What are the salient differences among the three major lines of descent?

In addition, the problem of evolution of metabolism needs to be addressed. What was the nature of pregenetic "metabolism"? Was it a basically dark reaction, solution biochemistry as is commonly believed? Or was it basically a "membrane" (surface) chemistry? Although pregenetic, was it nevertheless cellular? Did a primitive, pregenetic metabolic network develop any refining quality; i.e., did it tend to organize itself, become restrictive, more specific? What were the primitive catalysts? Reaction centers? How do they relate to today's co-enzymes and prosthetic groups? What was the relationship, the transition, between pregenetic "metabolism" and what would later be the true cellular metabolism? It has been customary to think that the first cells were heterotrophs and so had very little metabolism of their own (they took all amino acids, nucleotides, etc., from their rich growth medium). Is this a correct view? Is the Horowitz hypothesis for the origin of metabolic pathways (i.e., by backward evolution, one enzymatic step at a time) a correct view?

Also, we must be concerned with the problem of primitive energy sources. What mechanism generated chemical energy for prebiotic, pregenetic systems? Was there a pregenetic photosynthesis (visible-infrared range)? To what extent were chemical reactions, such as $CO_2 + 2H_2 \rightarrow CH_4 + 2H_2O$, present and utilized? Did membrane-associated energy production occur, such as through systems that automatically generated transmembrane H^+ (or other) gradients?

We should, in addition, elucidate the nature of the development of a genetic (informational) system. What is the molecular mechanism of translation? How do various versions (eucaryotic, archeabacterial, eubacterial) of the ribosome differ from one another, and what does this tell us about the origin of that structure? What is the most primitive form of the translation apparatus? How did the genetic code evolve? What is the relationship between its evolution and the various stages in the evolution of the translation mechanism? What is the relationship between nucleic-acid replication (or its transcriptions) and translation? What is the significance of the fact that bacterial RNA viral replicase has four subunits, three of which are associated with the translation process in the host cell? What was the nature of the aboriginal genome? Was it RNA or DNA? How was it organized? What were the aboriginal genes (what functions were encoded)? What was the relationship between the aboriginal genes and their gene products?

Finally, we must investigate the nature of the eucaryotic cell. How did it evolve? To what extent and in what ways are endosymbioses responsible for the uniqueness of the eucaryotic cell? Was the eucaryotic cell basically formed by the fusion of various procaryotes? Or, did most of the important characteristics that are specifically eucaryotic stem from a pre-procaryotic stage in evolution? To what extent and in what ways (if any) is the eucaryotic cell poly-phyletic in origin? More broadly, to what extent is it chimeric? How many (and what) major lines of eucaryotes are there? (In other words, how many kingdoms lie hidden in the general classification protista?) This question is underlain by the important general one of how easily various major states in evolution are arrived at. The wide range of questions raised points to a future growth of knowledge in this area. The origin of the cell may lay hidden in the biological record.

To quote the biochemist, Szent-Gyorgyi: "Life has developed its processes gradually, never rejecting what it has built, but building over what has already taken place. As a result the cell resembles the site of an archeological excavation with the successive strata on top of one another, the oldest one the deepest. The older a process, the more basic a role it plays and the stronger it will be anchored, the newest processes being dispensed with most easily."

CHEMICAL EVOLUTION

As we have pointed out in chapter V, the field of chemical evo-lution has been guided by the premise that the primitive atmosphere was hydrogen-rich, a reducing atmosphere with the major forms of carbon and nitrogen being methane and ammonia. However, recent models of the Earth's early atmosphere have placed severe limitations on the amount of hydrogen originally present. Since such a primitive atmosphere would be dominated by carbon dioxide and nitrogen, the capability of this atmosphere to sustain synthesis of amino acids and other biomonomers might be limited. The possibility that there was no soup of any complexity must therefore be considered and explored experimentally.

Thus, the field of the synthesis of biomonomers must be broadened to encompass not only experiments as they have been classically conceived (i.e., in the reducing atmosphere of the Miller-Urey model) but also, experiments utilizing nonreducing gases (i.e., CO, CO_2, N_2, etc.), minerals, and light. To the above must be added the constraints imposed by our expanding knowledge of the early Earth.

These research areas overlap the research interests of scientists in the solar-energy conversion field who are actively exploring metal surfaces and minerals which, in the presence of light, will photolyze water and reduce carbon dioxide and nitrogen. Therefore collaboration in this area should be encouraged between prebiotic chemists and scientists attempting to capture solar energy by chemical means.

Recent studies suggest clays, mineral surfaces, and metal ions had important roles in activating organic molecules for reaction, and stabilizing polymers by binding them to surfaces and then catalyzing condensation reactions. The oxidation state of the metal ion used is critically important and it should be compatible with the expected oxidation level of the metal ion in the presence of the reducing environment of the primitive Earth. Further research on catalysis and absorption by inorganic substances may help explain why only a limited group of organic molecules were included in living systems.

Many compounds with UV and visible chromophores are produced in experiments which simulate chemical transformations on the primitive Earth. In only a few instances the role of UV light in the further transformation of these compounds has been investigated. Since light in the UV-visible range was one of the most potent energy sources impinging on the primitive Earth, this area of research merits more extensive investigation. Of particular importance is the possibility of utilizing the light energy to drive reactions that would normally be energetically unfavorable, such as peptide or nucleotide bond formation.

It was recently observed that the sulfate in sea water is reduced to sulfide when it comes in contact with molten portions of the Earth's crust at the Galapagos Rift. This sulfide is emitted as H_2S or metal sulfides from these thermal vents. The ease of reduction of sulfate suggests that most of the sulfur on the primitive Earth was in the sulfide oxidation state. The role of sulfide ion and insoluble metal sulfides in chemical evolution deserves more detailed investigation. The pronounced (nucleophilic) reactivity of the sulfide ion could result in a marked change in the reaction pathways observed in its absence.

A case for a role of HCN in chemical evolution has been established by previous research. Purines, pyrimidines, and amino acids have been synthesized from HCN. Further analysis of the monomeric building blocks formed from HCN merits investigation. This study should include the investigation of catalysis by mineral surfaces and metal ions, since little has been done to study the complex interaction between organic and inorganic species on the putative primitive Earth.

Apart from the fashionable areas of amino acids and bases, the synthesis of other biomonomers should be pursued. For example, is the "formose" reaction the correct answer to the prebiotic formation of the more important sugar compounds? Ribose is formed in low yield by the base-catalyzed condensation of formaldehyde. A complicated mixture of C_4, C_5, C_6, and C_7 branched and linear sugars are made in this reaction. Either ribose was formed under a more unique set of reaction conditions or else there was a mechanism for the selection of ribose from this complex mixture. In addition, we still do not have prebiotic syntheses for long chain unbranched fatty acids, alcohols, aldehydes, or isoprenes with relatively high yields.

In the near future we can anticipate more work on membranes, and two specific questions need to be resolved. How and when did specific transport systems for solutes through lipid bilayers arise? Can a plausible protocell be built from amphipathic lipids, proteins, and (pre)nucleic acids? We anticipate a large input into prebiotic chemistry from the experimental study of lipid bilayers, and research in this area should be encouraged.

An important adjunct to the question of the formation of biomonomers is the origin of optical asymmetry. For example, was the selection of L-amino acids as the constituents of proteins a matter of chance, or was it the result of some asymmetric process on the preprebiotic Earth? Further work in these areas is warranted.

In the last few years an abiotic synthesis of glycine has been suggested as occurring in a spreading rift zone, specifically in the Red Sea. This suggests that the submarine hydrothermal systems should be studied as a possible model system for prebiotic chemistry.

Environments on the Earth are subject to all kinds of fluctuations: diurnal, seasonal, and tidal. For example, in some experiments environmental fluctuations of temperature and moisture content have been successfully used to produce peptides from amino acids. Overall, how important were fluctuations for organic chemical evolution? Were they necessary? These are but two questions we can ask relative to this potentially important regime.

Most experiments that model events on the primitive Earth focus on one or two steps in the process of chemical evolution. With the advent of sensitive analytical techniques it is now feasible to set up long term experiments where reactants are gradually added and removed from a flask over a long period of time in a situation which mimics the formation and reaction of biomolecules on the early Earth. The effect of changing the temperature, the exposure to light, and the reactants can be followed over periods of weeks and months by analysis. This experimental approach provides a more accurate model of the flux of chemicals through the primitive oceans and hence should provide useful information concerning the rates at which specific biomolecules were formed. This approach would also be ideal for the investigation of reactions taking place using mixtures of several reactants. For example, the reaction of carbon dioxide, hydrogen cyanide, formaldehyde, and UV light could be investigated in one such system.

As has been pointed out, much clearly remains to be learned about monomers. If, however, we assume for the purpose of the following discussion that, in principle, the problem of monomer production is solved, the next step requires condensation of amino acids into polypeptides and of bases, sugars, and phosphate into mono- and poly-nucleotides. This problem is still unsolved, despite the fact that many successful condensations have been carried out utilizing, as reagents, products of the electric-discharge reaction. The reason this problem remains unsolved is that, for ease of experimentation, only a limited number of organic molecules are used in most of these condensation studies. For example, amino acids (used in experiments to produce peptides) are only a fraction of the total organic compounds produced in the spark discharge, and amino acids would be expected to react with non-amino acids most of the time if the complex mixture were heated. The same criticism can be made of most of the "prebiotic" dehydration condensations that have been published.

Unless some experiment using complex mixtures of monomers actually yields polymers of interest, it would appear that a process of fractionation must have intervened between the formation of the original prebiotic pool of organic compounds and their condensation into biologically useful polymers. Some related questions include whether amino acids and nucleic acids can polymerize in the same system without interfering with each other. Also, can relevant reactions carried out with optically pure starting materials be done with racemic starting materials? The study of pure systems is certainly easier, but the study of mixed systems should be encouraged due to the important unanswered questions in this area. These questions are among the more important ones facing workers in this field at the present time, since they bear on the order of events in the origin of life; e.g., whether polypeptides or polynucleotides appeared first, as well as whether laboratory experiments are relevant to more complex situations on the prebiotic Earth.

MODELS FOR EARLY LIFE FORMS

Since we define life in terms of its genetic properties, and since the only known system possessing these properties is the protein-nucleic acid system, the most easily defended position holds that the first living things were based on this system. However, the spontaneous origin of such a complex mechanism poses great conceptual difficulties. Therefore, other possibilities should be considered. An important constraint is that the original self-replicating system, whatever it may have been, must have had the capability of evolving into the protein-nucleic acid system. Possibilities that are worth exporing include: (1) polynucleotides with some catalytic capability and (2) polypeptides with some replicative capability. Obviously, these systems would be extremely inefficient in comparison with the highly evolved modern mechanism, but they might have been sufficiently accurate to survive and evolve under the benign conditions of the primordial Earth. Either one might have been capable of developing into the modern cell.

The question of whether polynucleotides alone can constitute a self-replicating system can in part be answered by the development of experimental models for nonenzymatic replication. Some progress has been made in understanding how preformed polynucleotide chains can function as templates to direct the synthesis of their complements in nonenzymatic reactions. However, much remains to be learned about the catalytic effect of peptides, metal ions, etc.

Work on the incorporation of pyrimidines in template-directed reactions is also important. Further study of template-directed reactions should contribute to our understanding of the origins of nucleic acid replication.

The minimal living system must be self-duplicating and mutable, and it must have, at least latently, the capacity for heterocatalysis for bringing about chemical changes in the environment that support the self-duplication function. In other words, a living thing must be capable of rearranging the universe to produce more of itself; and the "self" must be capable of continual change. All of this implies information storage, replication, retrieval, and utilization. Contemporary life displays these properties in a combined protein/nucleic acid system, so further work in this area is especially desirable. The major unresolved problem is in the area of translation and genetic coding which features the specific interactions of amino acids with nucleic acids or short oligonucleotides, and formation of peptide bonds by template mechanisms. Both are worth pursuing. The demonstration of translation in an experimental model that combines both of the above features is of the greatest importance. A wide variety of experimental approaches is justified in the absence of a consensus on precise mechanisms.

The study of simple peptides or polypeptides as catalysts is important in understanding how a genetic system, once started, could have gained a selective advantage. The processes of replication and translation might be more efficient or more selective when catalyzed by protoenzymes rather than without them. Case models which leave out catalysts will not tell the whole story. Since the synthesis of molecules resembling proteins and nucleic acids requires removal of water, it is not surprising that molecules which have been studied as catalysts often lead to degradation rather than synthesis when the reaction occurs in water solution. The study of condensed phases, mineral surfaces, membranes, and other heterogeneous systems may be more conducive to synthetic catalysis than the study of the homogeneous aqueous phase.

THE REPLICATION OF SURFACE PATTERNS – AN
ALTERNATIVE ENTRY FOR LIFE

The identification of the large biopolymers – proteins and nucleic acids – as the significant agents in self-replication has been the motif of most of this volume. The implication taken is plain – the earliest examples of such a well-developed system as we see at work universally in life today must have been in some way simpler analogies of the two classes of substances. The essentially homogeneous nature of the self-replicating preparations of enzyme, energy source, monomers, and template in water suspension has been seen as an advanced version of the original natural system. That beginning system would have been a simpler, less adapted set of much smaller molecules in the same sort of solution where the chemistry is basically similar to that of the modern examples.

It is clear that the inference, suggestive and powerful though it is, is not unique. The gap in time allows the postulation of other distinct systems of entry that are discontinuous with the present informational but continuous in metabolic mechanisms, and which might step by step have come to develop the powerful self-replication that is universal today. For example, very recent experimental support has been found for a suggestion, 15 years old or more, that differs in most ways from the prototype of a linkage between the two classes of polymers. This hypothesis suggests replicating molecular patterns in two dimensions from a solid surface and not in one dimension from a long helical polymer. It links an inorganic substrate in an essential way with the organic product. It seems prudent to recognize that neither of these two incomplete models is apt to contain the whole story, but the whole domain of simple models spans between them as extremes. Consideration for both of the poles is surely the wisest strategy. In the next pages we present a summary of the model that rests on an inorganic notion of the first steps toward life.

Monomer Supply and the Early Atmosphere

The organic chemists view the decisive polymers as links of the very common atoms, C, H, N, O, P, and S. The two essential heavier atoms, S in proteins and P in the nucleic acids, are full partners of the more usual atoms of organic molecules. The most abundant mineral compounds of the planetary surface contain as well, atoms still more abundant than are P and S, but not found at all within the great polymers. Among these are in particular magnesium, silicon, iron, and the sources of the ions of natural waters, especially Na and K. These inorganic elements are involved in life today in a more or less essential way, though they are not part either of protein or of nucleic acid. The inorganic geochemists take these atoms as central to their studies just as the biochemists look on the others as dominant. Their possible interaction certainly should not be overlooked simply because it does not quite lie within the most-studied disciplines of carbon chemistry.

As pointed out in chapter V, the reduction both of carbon dioxide and nitrogen are possible on a primitive Earth, once the right electron donors (e.g., Fe^{++}) are present. The involvement of light to drive the reactions is at least reminiscent of photosynthesis. These photosynthetic models should be further explored.

Replication

As we have previously defined, a minimal living system capable of evolution must be self-duplicating and mutable; it must have, at least latently, the capacity for heterocatalysis. An example of such a minimal definition has recently been met by an experiment in which surface charge patterns on clay particles in water were shown to replicate and mutate. The preliminary results of these experiments indicate that clay minerals such as montmorillonite, which can swell to a large degree, may be capable of replicative self-multiplication. These minerals may, therefore, be looked upon as models for proto-life, or possibly for most primitive life; their catalytic capabilities and selectivities can be altered, and thus can accelerate or retard the rate of self-multiplication.

Results like those were foreseen by Cairns-Smith in 1965. He suggested that the primitive genes were patterns of substitutions in colloidal clay crystallites. The theoretical information density in such crystallites is comparable to that in DNA. Evolution proceeded through selective elaboration of pattern mutations that had survival value for the clay crystallites that held them.

If we take it as a fact that such clay-surface charge patterns are replicable, and that changes in them can induce stable changes in their progeny, it is hard to deny the self-replicating quality. Notice one very substantial difference from the biopolymer model: replication in dimension is different. The clay particles replicate two-dimensional patterns (the laboratory examples show that some 10^6 ionic sites are replicated) while the nucleic acids instead replicate a linear sequence that is one dimensional.

Naturally there is a very long path from such curious "living" mineral particles to the cells we ascribe to or even find in the fossil stromatolites. The gap can be closed only by hard work in the laboratory and by new ideas. But it is interesting to put forward an optimistic, if vague, scenario about how the scheme might have gone forward in the gap between 4.5 and 3.5 or 3.8 b.y. ago. Of course, this is meant only as a pedagogical example.

One such scenario would go as follows. Once the early Earth had well-differentiated into core, mantle, and crust, the atmosphere would be mainly carbon dioxide, nitrogen, and water, with some minor constituents like hydrogen sulfide. This atmosphere would not give rise to a soup of monomers, even locally. The interaction between atmosphere, sea water, and the silicious, iron-rich crust of the Earth, would lead rather to copious formation of clays. The iron-rich clays replicated during many cycles of inundation and dryness, mutated, and began to fix carbon dioxide photochemically, using solar UV and ferrous ion. This could lead to sugars, to the citric-acid cycle, and even to fatty acids. In a later stage, the fixation of molecular nitrogen occurred as well, and the surface formation of amino acids and nucleotides became possible. The evolving clays began to polymerize these surface monomers. In this system, the nucleic acids became coupled to the polypeptides through a genetic code. From this complex surface-borne system, a newly self-enclosed system based on nucleic acids and proteins began a new and independent evolution of its own, free of solid substrate — a protocell with its membrane consisting of some lipid-rich layers.

It is evident that in such ideas we have the beginnings of a rich and promising experimental campaign which is complementary to the search for coupling between the simpler biopolymers held in solution.

SUMMARY

What is the simplest chemical system that is capable of complete genetic self-replication and open-ended Darwinian evolution? This question has been at the heart of our discussion of the origin of life. Experiments designed to demonstrate true self-replication and natural selection in prebiotic situations, although perhaps a distant goal, are of obvious value. Three experimental systems which might conceivably lead to such a demonstration are nonenzymatic nucleic-acid replication, clay replication, and combined peptide/nucleic acid systems. Although none of these systems has demonstrated all of the characteristics for self-replication, each has promise worth pursuing.

Another possible route to demonstrate self-replication and natural selection, although without addressing the origin of such a system, is to try to construct a minimal self-replicating system from components of biological cells. This would indeed be worthwhile if it could be done, since it would to some extent plug the huge conceptual gap between simple self-replicating systems and the complex genetic system of even the simplest contemporary cell.

In summary, the most critical need in the areas of nucleic-acid replication, translation, and other aspects of self-replicating systems is for experimental studies rather than more speculations and diffuse theories.

NEXT ORGANIZATIONAL STEPS

This whole report has sought to sum up the present state of our knowledge and the questions which remain, the many small questions whose answers will lead us to see into the great question, so important for most reflective people, scientists or not.

A modest world community — a few scores of laboratories and a thousand or two scientific workers — will encounter great difficulty in its pursuit of this very important question. People in such a community work with a given discipline; they are geochemists, or microbiologists or nucleic acid chemists, or biochemists, or specialists in planetary dynamics. The classes they teach, the techniques they use, the meetings they attend, the journals they read and contribute to, are typically diverse. The normal structure of science, especially university science, is disciplinary. Unlike the mission-oriented inter-disciplinary teams of NASA, scientific experiment tends to be discipline-bound. A journal on the Origin of Life is exceptional since it centers on a large question, not at all typical of the journals in which most work must be published.

We believe it urgent that some effort be exerted to strengthen the interdisciplinary nature of this work, which we have character-ized as seeking the answer to a large question, not merely working out the answers to many small questions, though that is, of course, the indispensable path to most progress.

To this end we offer two recommendations, based on discussion and experience within this widely interdisciplinary workshop:

1. There is no way to make continued progress unless young, talented research workers are steadily recruited to the work. But in the absence of a well-established discipline, young people are much more likely to seek surer and easier paths. A direct incentive should be provided, an incentive which would enlarge the opportunities for young people who wished to undertake some portion of this great question as their own work.

We recommend that some funding agency — a joint effort of several, whether Federal or private — undertake to offer a yearly grant of some post-doctoral Fellowships in the Origin of Life. They should be grants for 2 or 3 years, perhaps, tenable at any place which has agreed to accept the person. Possibly, the grant should include not only a reasonable sum for salary and travel, but also some addi-tional funds to encourage the host laboratory to accept the Fellow. The scale of the grants would of course depend on funds available; a substantial effect could be achieved by the U.S.A. by grants say to 10 persons a year, as a steady-state number. The competition should be open to research people from any discipline and any country; the only requirement would be a showing of the relevance and hope of the study for progress towards a knowledge of the origins of life.

It would be most appropriate to call these Fellowships by the name of the late Harold Urey; he was the remarkable American scien-tist who as much as any other man has opened the field to modern study. Perhaps that name would open some new sources of funding;

a memorial so constructed would be fitting in the highest degree to the memory of Harold Urey.

2. Such a scheme helps meet the fundamental task of a long-range research program — careers for young investigators, given form within an interdisciplinary framework. There is another valuable device for so broad a field. It is not a response to a steady problem; rather, it is a means of bringing resources to bear on opportunities for great progress as they arise. (By good fortune, we saw it at work at the University of California, Los Angeles (UCLA), where a group organized by Dr. William Schopf out of a windfall award from the National Science Foundation (NSF) studies the Precambrian record.) The idea is simple and attractive. From time to time, set not by the calendar but by the state of knowledge, a group of research people of differing skills and approaches can be brought together to spend a limited time as a team. They would probably be housed at one central laboratory, bringing with them expertise, even equipment, that they already possess. They work jointly for a while, yet from differing specialties, at a complex of problems they recognize as ripe for a joint attack. One might call this a Focus Award in the Origin of Life. It would be given to any investigator who would persuade the reviewers that the time, place, and people were right for say a 2-year joint effort by from four to ten investigators. The Focus might link paleontology to biochemistry or astronomy, or it could span even wider disciplines. The award would not be made each year, or in any other routine way, but only on the showing that the moment had arrived to strengthen the dispersed and diverse research in the broad field by setting up a team, not for a long career, but for a limited time. Once every few years it seems likely that such a Focus will come to make sense, at a level that is overall small compared to the steady flow of support; it is obviously a very attractive but quite uncertain program. We have not addressed ourselves to the administrative problems of making such awards and making certain of fiscal responsibility; the task is not easy, but surely soluble. Our view is that the scientific merit of the proposal and the personal reputation of the proposers would have to be so high that difficulties can be overcome with relative ease. The point is to supply this important but diffuse field with the chance for the sort of concentration that bigger laboratories with well-defined missions can now direct at their

problems. At the same time we do not propose new organizations, but rather to allow serious temporary cooperation at a level of real effectiveness. The teams ought not to be too small, nor too large; the tasks neither too brief nor too extensive.

With a steady Urey Fellowship Program in the Origin of Life and the tempting opportunity of a Focus Award in the Origin of Life we feel that the tasks we have outlined, tasks which transcend any discipline of science but which promise answers of the highest importance to one of the deepest questions human beings can ask, can be met in the decade or so ahead. Unless some such new support is found outside the normal rubrics, there will not be much progress toward the solution of questions too profound for chemists, biologists, astronomers, or geologists to answer alone in the ordinary flow of the stream of contemporary science. Like all real science, this fundamental investigation also has foreseeable applications. For do not our fossil fuels represent ancient processes of organic chemistry, not deeply understood? And does not the ecology of the life in the shallow waters bear sharply on the great chemical cycles that can fit or spoil the Earth for human life? These are mere accidental, but urgent by-products of a deep study of early life and its nature. We thus end this book with a note of hope that these suggestions attest to the vigor of this field. The depth and variety of questions to be answered are signs of the maturation of this exciting scientific endeavor. If we are wise enough in this small research investment, we can expect real advances in knowledge and in practice of facing the great questions that reflective people ask: what is life, and how did it arise within the context of changing nature?

APPENDIX

MOLECULAR STRUCTURES
OF IMPORTANT BIOLOGICAL COMPOUNDS

The Twenty Amino Acids Found in Proteins

GLYCINE (gly)

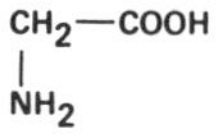

$$CH_2-COOH$$
$$|$$
$$NH_2$$

ALANINE (ala)

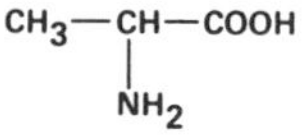

$$CH_3-CH-COOH$$
$$|$$
$$NH_2$$

VALINE (val)

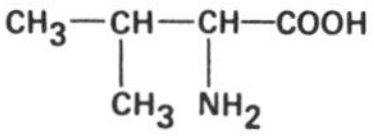

$$CH_3-CH-CH-COOH$$
$$|\ \ \ |$$
$$CH_3\ NH_2$$

ISOLEUCINE (ile)

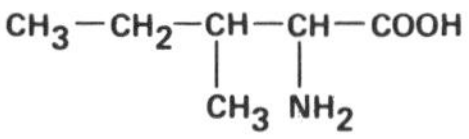

$$CH_3-CH_2-CH-CH-COOH$$
$$|\ \ \ |$$
$$CH_3\ NH_2$$

LEUCINE (leu)

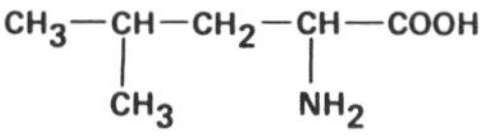

$$CH_3-CH-CH_2-CH-COOH$$
$$|\ \ \ \ \ \ \ \ \ |$$
$$CH_3\ \ \ \ \ \ \ NH_2$$

PROLINE (pro)

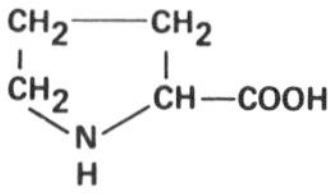

$$CH_2-\!-\!-CH_2$$
$$|\ \ \ \ \ \ \ \ \ |$$
$$CH_2\ \ \ \ \ \ CH-COOH$$
$$\diagdown\ N\ \diagup$$
$$H$$

SERINE (ser)

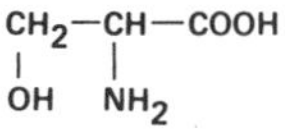

$$CH_2-CH-COOH$$
$$|\ \ \ \ |$$
$$OH\ \ NH_2$$

THREONINE (thr)

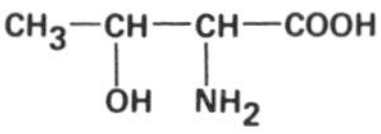

$$CH_3-CH-CH-COOH$$
$$|\ \ \ \ |$$
$$OH\ \ NH_2$$

ASPARTIC ACID (asp)

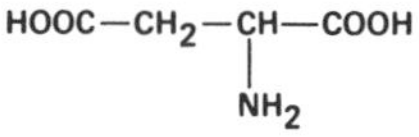

$$HOOC-CH_2-CH-COOH$$
$$|$$
$$NH_2$$

GLUTAMIC ACID (glu)

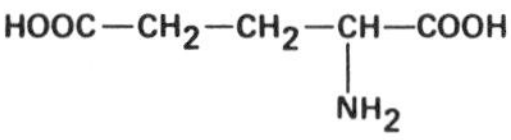

$$HOOC-CH_2-CH_2-CH-COOH$$
$$|$$
$$NH_2$$

CYSTEINE (cys)

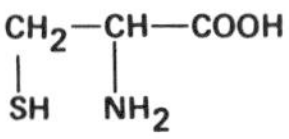

$$CH_2-CH-COOH$$
$$|\ \ \ \ |$$
$$SH\ \ NH_2$$

METHIONINE (met)

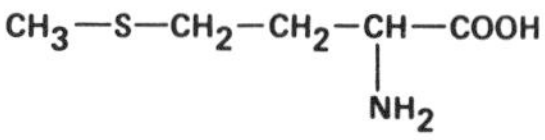

$$CH_3-S-CH_2-CH_2-CH-COOH$$
$$|$$
$$NH_2$$

The Twenty Amino Acids Found in Proteins (continued)

ASPARAGINE (asn)

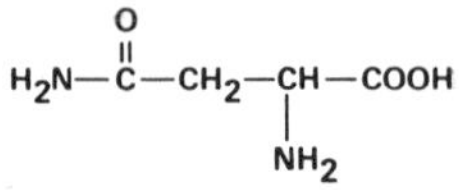

$$H_2N-\overset{\overset{\text{O}}{\|}}{C}-CH_2-CH-COOH$$
$$|$$
$$NH_2$$

GLUTAMINE (gln)

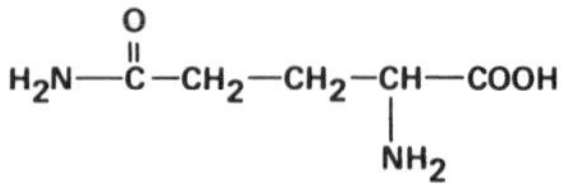

$$H_2N-\overset{\overset{\text{O}}{\|}}{C}-CH_2-CH_2-CH-COOH$$
$$|$$
$$NH_2$$

ARGININE (arg)

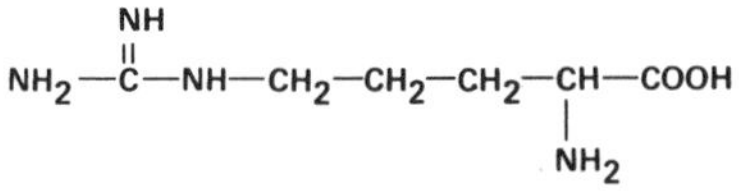

$$NH_2-\overset{\overset{\text{NH}}{\|}}{C}-NH-CH_2-CH_2-CH_2-CH-COOH$$
$$|$$
$$NH_2$$

LYSINE (lys)

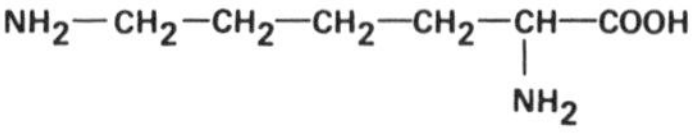

$$NH_2-CH_2-CH_2-CH_2-CH_2-CH-COOH$$
$$|$$
$$NH_2$$

HISTIDINE (his)

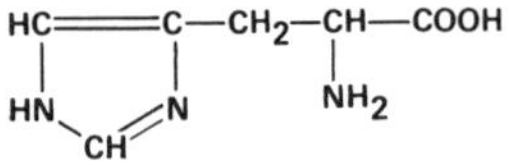

$$HC=C-CH_2-CH-COOH$$
$$| \quad | \qquad |$$
$$HN \quad N \quad NH_2$$
$$\diagdown CH \diagup$$

PHENYLALANINE (phe)

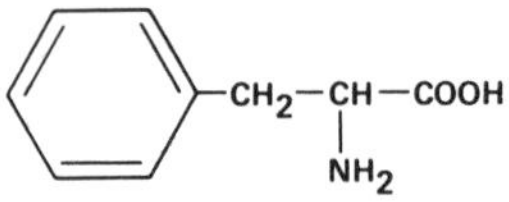

$$\text{⬡}-CH_2-CH-COOH$$
$$|$$
$$NH_2$$

TYROSINE (tyr)

$$HO-\text{⬡}-CH_2-CH-COOH$$
$$|$$
$$NH_2$$

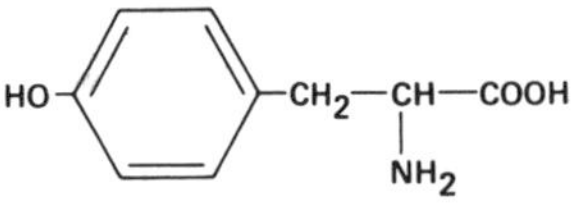

TRYPTOPHAN (trp)

$$-CH_2-CH-COOH$$
$$|$$
$$NH_2$$

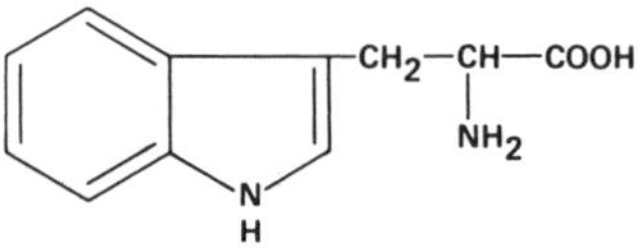

The Four Nucleotides Found in DNA

INDEX